NATURAL DISASTER RESEARCH, PREDICTION AND MITIGATION

DEEP-WELL INJECTIONS AND INDUCED SEISMICITY

UNDERSTANDING THE RELATIONSHIP (WITH ACCOMPANYING CD-ROM)

NATURAL DISASTER RESEARCH, PREDICTION AND MITIGATION

Additional books in this series can be found on Nova's website under the Series tab.

Additional e-books in this series can be found on Nova's website under the e-book tab.

NATURAL DISASTER RESEARCH, PREDICTION AND MITIGATION

DEEP-WELL INJECTIONS AND INDUCED SEISMICITY

UNDERSTANDING THE RELATIONSHIP (WITH ACCOMPANYING CD-ROM)

JAMES B. FERNANDEZ
EDITOR

nova publishers
New York

Library of Congress Cataloging-in-Publication Data

ISBN: 978-1-63482-557-3

Published by Nova Science Publishers, Inc. † New York

CONTENTS

PREFACE

This purpose of this book is to discuss the relationship between deep-well injections and induced seismicity.

Chapter 1 - The development of unconventional oil and natural gas resources using horizontal drilling and hydraulic fracturing (fracking) has created new demand for wastewater disposal wells that inject waste fluids into deep geologic strata. An increasing concern in the United States is that injection of these fluids may be responsible for increasing rates of seismic activity. The number of earthquakes of magnitude 3.0 or greater in the central and eastern United States has increased dramatically since about 2009, from an average of approximately 20 per year between 1970 and 2000 to over 100 per year in the period 2010-2013. Some of these earthquakes may be felt at the surface. For example, 20 earthquakes of magnitudes 4.0 to 4.8 have struck central Oklahoma since 2009. The largest earthquake in Oklahoma history (magnitude 5.6) occurred on November 5, 2011, near Prague, causing damage to several structures nearby. Central and northern Oklahoma were seismically active regions before the recent increase in the volume of waste fluid injection through deep wells. However, the recent earthquake swarm does not seem to be due to typical, random, changes in the rate of seismicity, according to the U.S. Geological Survey.

The relationship between earthquake activity and the timing of injection, the amount and rate of fluid injected, and other factors are still uncertain and are current research topics. Despite increasing evidence linking some deep-well disposal activities with human-induced earthquakes, only a small fraction of the more than 30,000 U.S. wastewater disposal wells appears to be associated with damaging earthquakes.

The potential for damaging earthquakes caused by hydraulic fracturing itself, as opposed to deep-well injection of wastewater from oil and gas activities, appears to be much smaller. Hydraulic fracturing intentionally creates fractures in rocks, and induces microseismicity, mostly of less than magnitude 1.0, too small to feel or cause damage. In a few cases, however, fracking has led directly to earthquakes larger than magnitude 2.0, including at sites in Oklahoma, Ohio, England, and Canada.

The Environmental Protection Agency's (EPA's) Underground Injection Control (UIC) program under the Safe Drinking Water Act (SDWA) regulates the subsurface injection of fluids to prevent endangerment of drinking water sources. EPA has established regulations for six classes of injection wells, including Class II wells used for the injection of fluids for enhanced oil and gas recovery and wastewater disposal. Most oil and gas states administer the UIC Class II program. The SDWA does not address seismicity, although EPA regulations for certain classes of injection wells require some evaluation of seismic risk. Such requirements do not apply to Class II wells; however, EPA has developed a framework for evaluating seismic risk when reviewing Class II permit applications in states where EPA administers this program. How Congress shapes EPA or other agency efforts to address and possibly mitigate human-caused earthquakes may be an issue in the 114th Congress.

In 2011, in response to seismic events in Arkansas and Texas thought to be associated with wastewater disposal wells, EPA authorized a national UIC technical work group to develop recommendations to address the risk of Class II disposal-induced seismicity. EPA plans to issue a document outlining technical recommendations and best practices in early 2015. At the state level, several states have increased oversight of Class II wells in response to induced seismicity concerns. In 2014, state oil and gas and groundwater protection agencies established a work group to discuss Class II disposal wells and recent seismic events occurring in multiple states.

Chapter 2 - The U.S. Environmental Protection Agency (EPA) Underground Injection Control (UIC) program regulates injection of fluids related to oil and gas production as Class II injection wells for the protection of underground sources of drinking water (USDWs). Because seismic events from injection have the potential to cause endangerment of underground sources of drinking water, the UIC program director should be aware of that potential and be prepared with response options should seismic events become a concern. Unconventional resources and new technologies, such as horizontal drilling and advanced completion techniques, have expanded the geographic area for oil and gas production activities, resulting in a need for Class II disposal wells in some areas previously considered unproductive.

Recently, a number of small to moderate magnitude (M<5.0) earthquakes have been recorded in areas with Class II disposal wells related to shale hydrocarbon production. To address the concern that induced seismicity could interfere with containment of injected fluids and endanger drinking water sources, EPA's Drinking Water Protection Division requested that the UIC National Technical Workgroup (NTW) develop a report with practical tools to help UIC regulators address injection-induced seismicity. The Induced Seismicity Working Group (WG) of the NTW developed this report in response, using the existing Class II regulatory framework to provide possible strategies for managing and minimizing the potential for significant injection-induced seismic events. The report focuses on Class II disposal operations and not enhanced recovery wells or hydraulically fractured wells. In formulating the strategies in this report, the NTW conducted a technical literature search and review. Additionally, the NTW evaluated four case examples (in Arkansas, Ohio, Texas, and West Virginia) and considered data availability and variations in geology and reservoir characteristics.

Chapter 3 - Portions of all 50 states and the District of Columbia are vulnerable to earthquake hazards, although risks vary greatly across the country and within individual states. Seismic hazards are greatest in the western United States, particularly in California, Washington, Oregon, and Alaska and Hawaii. California has more citizens and infrastructure at risk than any other state because of the state's frequent seismic activity combined with its large population and developed infrastructure.

The United States faces the possibility of large economic losses from earthquake-damaged buildings and infrastructure. The Federal Emergency Management Agency has estimated that earthquakes cost the United States, on average, over $5 billion per year. California, Oregon, and Washington account for nearly $4.1 billion (77%) of the U.S. total estimated average annualized loss. California alone accounts for most of the estimated annualized earthquake losses for the nation.

A single large earthquake, however, can cause far more damage than the average annual estimate. The 1994 Northridge (CA) earthquake caused as much as $26 billion (in 2005 dollars) in damage and was one of the costliest natural disasters to strike the United States. One study of the damage caused by a hypothetical magnitude 7.8 earthquake along the San Andreas Fault in southern California projected as many as 1,800 fatalities and more than $200 billion in economic losses.

Unlike other natural hazards, such as hurricanes, where predicting the location and timing of landfall is becoming increasingly accurate, the scientific understanding of earthquakes does not yet allow for precise earthquake prediction. Instead, notification and warning typically involve communicating the location and magnitude of an earthquake as soon as possible after the event to emergency response providers and others who need the information.

A precise relationship between earthquake mitigation measures, federal earthquake-related activities such as earthquake research, and reduced losses from an actual earthquake may never be possible. However, as more accurate seismic hazard maps evolve, and as understanding of the relationship between ground motion and building safety improves, trends denoting the effectiveness of mitigation strategies and earthquake research and other activities may emerge more clearly. Without an ability to precisely predict earthquakes, Congress is likely to face an ongoing challenge in determining the most effective federal approach to increasing the nation's resilience to low-probability but high-impact major earthquakes.

In: Deep-Well Injections and Induced Seismicity ISBN: 978-1-63482-557-3
Editor: James B. Fernandez

Chapter 1

HUMAN-INDUCED EARTHQUAKES FROM DEEP-WELL INJECTION: A BRIEF OVERVIEW*

Peter Folger and Mary Tiemann

SUMMARY

The development of unconventional oil and natural gas resources using horizontal drilling and hydraulic fracturing (fracking) has created new demand for wastewater disposal wells that inject waste fluids into deep geologic strata. An increasing concern in the United States is that injection of these fluids may be responsible for increasing rates of seismic activity. The number of earthquakes of magnitude 3.0 or greater in the central and eastern United States has increased dramatically since about 2009, from an average of approximately 20 per year between 1970 and 2000 to over 100 per year in the period 2010-2013. Some of these earthquakes may be felt at the surface. For example, 20 earthquakes of magnitudes 4.0 to 4.8 have struck central Oklahoma since 2009. The largest earthquake in Oklahoma history (magnitude 5.6) occurred on November 5, 2011, near Prague, causing damage to several structures nearby. Central and northern Oklahoma were seismically active regions before the recent increase in the volume of waste fluid injection through

* This is an edited, reformatted and augmented version of a Congressional Research Service publication R43836, prepared for Members and Committees of Congress dated January 8, 2015.

deep wells. However, the recent earthquake swarm does not seem to be due to typical, random, changes in the rate of seismicity, according to the U.S. Geological Survey.

The relationship between earthquake activity and the timing of injection, the amount and rate of fluid injected, and other factors are still uncertain and are current research topics. Despite increasing evidence linking some deep-well disposal activities with human-induced earthquakes, only a small fraction of the more than 30,000 U.S. wastewater disposal wells appears to be associated with damaging earthquakes.

The potential for damaging earthquakes caused by hydraulic fracturing itself, as opposed to deep-well injection of wastewater from oil and gas activities, appears to be much smaller. Hydraulic fracturing intentionally creates fractures in rocks, and induces microseismicity, mostly of less than magnitude 1.0, too small to feel or cause damage. In a few cases, however, fracking has led directly to earthquakes larger than magnitude 2.0, including at sites in Oklahoma, Ohio, England, and Canada.

The Environmental Protection Agency's (EPA's) Underground Injection Control (UIC) program under the Safe Drinking Water Act (SDWA) regulates the subsurface injection of fluids to prevent endangerment of drinking water sources. EPA has established regulations for six classes of injection wells, including Class II wells used for the injection of fluids for enhanced oil and gas recovery and wastewater disposal. Most oil and gas states administer the UIC Class II program. The SDWA does not address seismicity, although EPA regulations for certain classes of injection wells require some evaluation of seismic risk. Such requirements do not apply to Class II wells; however, EPA has developed a framework for evaluating seismic risk when reviewing Class II permit applications in states where EPA administers this program. How Congress shapes EPA or other agency efforts to address and possibly mitigate human-caused earthquakes may be an issue in the 114th Congress.

In 2011, in response to seismic events in Arkansas and Texas thought to be associated with wastewater disposal wells, EPA authorized a national UIC technical work group to develop recommendations to address the risk of Class II disposal-induced seismicity. EPA plans to issue a document outlining technical recommendations and best practices in early 2015. At the state level, several states have increased oversight of Class II wells in response to induced seismicity concerns. In 2014, state oil and gas and groundwater protection agencies established a work group to discuss Class II disposal wells and recent seismic events occurring in multiple states.

INTRODUCTION

Human-induced earthquakes, also known as induced seismicity, are an increasing concern in regions of the United States where the produced fluids and wastewaters from oil and natural gas activities are being injected into the subsurface through deep disposal wells. The immediate concern is that injection of these fluids into underground formations may be responsible for damaging earthquakes in regions that typically do not experience much seismic activity. Induced seismicity has garnered increased attention because of the rapid development of unconventional oil and gas resources, in part due to the use of hydraulic fracturing (often referred to as fracking). It is important to distinguish between seismic activity possibly related to hydraulic fracturing itself and the possibility of human-induced earthquakes related to injecting fluids down disposal wells, which may not be located near where wells were fracked.

Human activities have long been known to have induced earthquakes in some instances: impoundment of reservoirs, surface and underground mining, withdrawal of fluids such as oil and gas, and injection of fluids into subsurface formations. With the increase in the use of horizontal drilling and hydraulic fracturing to extract oil and gas from shale, and the concomitant increase in the amount of fluids that are injected for high-volume hydraulic fracturing and for disposal, there are several indications of a link between the injected fluids and unusual seismic activity. *Figure 1* illustrates conceptually the processes of deep-well injection and the linkage to triggering earthquakes.

The principal seismic hazard that has emerged from the increased amount of oil and gas activity in the United States appears to be related to disposal of wastewater using deep-well injection in some regions of the country. For example, in a May 2, 2014, joint statement between the Oklahoma Geological Survey and the U.S. Geological Survey (USGS), researchers reported a 50% increase in the rate of earthquakes in Oklahoma since 2013.[1] A USGS analysis of the rising trend suggested that a likely contributing factor was deep-well injection of oil-and-gas-related wastewater.[2] But the relationship between earthquake activity and the timing of injection, the amount and rate of fluid injected, and other factors are still uncertain and are current research topics. A 2013 article that reviewed the current understanding of human-caused earthquakes noted that, of the more than 30,000 wastewater disposal wells classified by the Environmental Protection Agency (EPA) as Class II,[3] only a small fraction appears to be associated with damaging earthquakes.[4]

The potential for damaging earthquakes caused by hydraulic fracturing itself, as opposed to deep-well injection of wastewater from fracking and other oil and natural gas production, appears to be much smaller. The 2013 review article indicated that the vast majority of wells used for hydraulic fracturing itself cause microearthquakes—the results of fracturing the rock to extract natural gas—which are typically too small to be felt or cause damage at the surface. The 2013 review documented a few cases where fracking itself caused detectable earthquakes felt at the surface, but these were too small to cause damage.

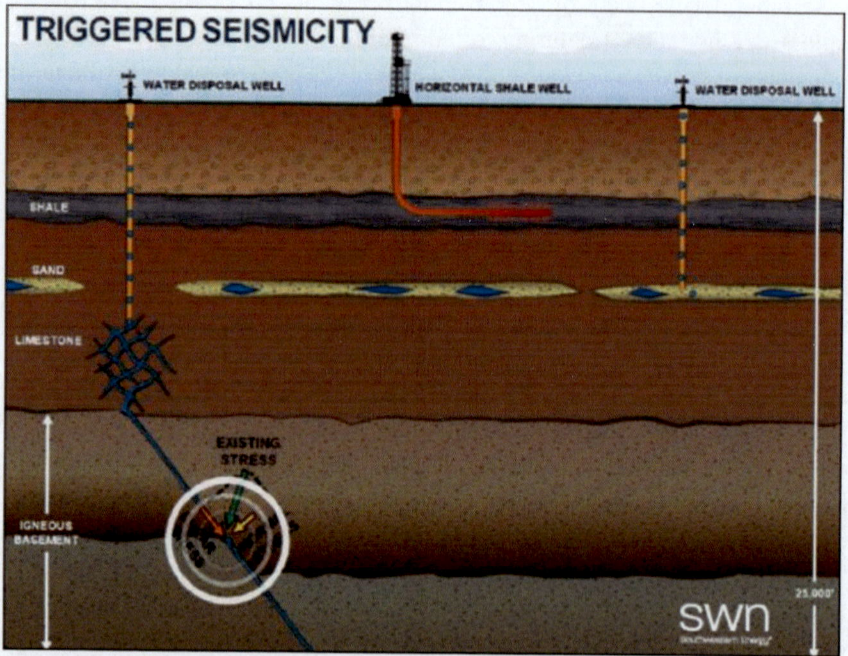

Source: North Carolina General Assembly, presentation by the Arkansas Oil and Gas Commission, *Fayetteville Shale Overview, for the North Carolina Delegation*, slide 33 prepared by Southwestern Energy, November 21, 2013, http://www.ncleg.net/documentsites/committees/BCCI-6576/2013-2014/5%20-%20 Feb.%204.%202014/Presentations%20and%20Handouts/Arkansas%20Site%20Visit%20Attachments/Att.%205%20-%20AOGC%20Presentation%2011-21-13%20%283%29.pdf.

Notes: The figure is for illustrative purposes only, and does not depict any specific location or geological formation.

Figure 1. Illustration of the Possible Relationship Between Deep-Well Injection and Induced Seismicity.

This report reviews the current scientific understanding of induced seismicity, primarily in the context of Class II oil and gas wastewater disposal wells. The report also outlines the regulatory framework for these injection wells, and identifies several federal and state initiatives responding to recent events of induced seismicity associated with Class II disposal.

Congressional Interest

How deep-well injection is linked to induced seismicity, and state and federal efforts to address that linkage, are of interest to Congress because of the implications to continued development of unconventional oil and gas resources in the United States. If the current boom in onshore oil and gas production continues, then deep-well injection of waste fluids is likely to also continue and may increase in volume. Also, what Congress, the federal government, and the states do to address and mitigate possible human-caused earthquakes from deep-well injection of oil and gas-related fluids may provide some guidance for the injection and sequestration of carbon dioxide. Carbon dioxide sequestration would involve ongoing, long-term, high-volume, high-pressure injection via deep wells. Several large-scale injection experiments are currently underway; however, the relationship between long-term and high-volume carbon dioxide injection and induced earthquakes is not known.

Current Scientific Understanding of Induced Seismicity in the United States

Since about the 1920s, it has been known that pumping fluids in and out of the Earth's subsurface has the potential to cause earthquakes.[5] In addition, a wide range of other human activities have been known to cause earthquakes, including the filling of large reservoirs, mining, geothermal energy extraction, and others.[6] The mechanics of how human industrial activities may cause earthquakes are fairly well known: the human perturbation changes the amount of stress in the earth's crust, and the forces that prevent faults from slipping become unequal. Once those forces are out of equilibrium, the fault ceases to be locked, and the fault slips, sending shock waves out from the fault that potentially reach the surface and are strong enough to be felt or cause damage.

Even knowing that human activities can cause earthquakes, and the mechanics of the process, it is currently nearly impossible to discriminate between man-made earthquakes and those caused by natural tectonic forces through the use of modern seismological methods.[7] Other lines of evidence are required to positively link human activities to earthquakes. That linkage is becoming increasingly well understood in parts of the United States where activities related to oil and gas extraction—deep-well injection of oil and gas wastewater, and hydraulic fracturing—have increased significantly in the last few years, particularly in Oklahoma, Texas, Arkansas, Ohio, Colorado, and several other states.[8] Nevertheless, the majority of these activities are not known to cause earthquakes; most are termed aseismic (i.e., not causing any appreciable seismic activity, at least for earthquakes greater than magnitude 3).[9] (See text box below for a brief description of earthquake magnitude and intensity.)

Earthquake Magnitude and Intensity[11]

Earthquake magnitude is a number that characterizes the relative size of an earthquake. It was historically reported using the Richter scale. Richter magnitude is calculated from the strongest seismic wave recorded from the earthquake, and is based on a logarithmic (base 10) scale: for each whole number increase in the Richter scale, the ground motion increases by 10 times. The amount of energy released per whole number increase, however, goes up by a factor of 32. The moment magnitude (M) scale is another expression of earthquake size, or energy released during an earthquake, that roughly corresponds to the Richter magnitude and is used by most seismologists because it more accurately describes the size of very large earthquakes. Sometimes earthquakes will be reported using qualitative terms, such as Great or Moderate. Generally, these terms refer to magnitudes as follows: Great (M>8); Major (M>7); Strong (M>6); Moderate (M>5); Light (M>4); Minor (M>3); and Micro (M<3). This report uses the moment magnitude scale, which is generally consistent with the Richter scale.[12]

Scientists currently have limited capability to predict human-caused earthquakes for a number of reasons, including uncertainty in knowing the state of stress in the Earth; rudimentary knowledge of how injected fluids flow underground after injection; poor knowledge of faults that could potentially slip and cause earthquakes; limited networks of seismometers (instruments used to measure seismicity) in regions of the country where most oil-and-gas-related activities are occurring; and difficulty in predicting how large an earthquake will grow once it is triggered.[10]

A Historical Example—The Rocky Mountain Arsenal

Prior to the moment magnitude (M) 5.6 earthquake that occurred on November 6, 2011, in central Oklahoma (discussed below), an M 4.8 earthquake that struck northeast Denver on August 9, 1967, was generally accepted as the largest recorded human-induced earthquake. The M 4.8 earthquake was part of a series of earthquakes that began within several months of the 1961 start of deep-well injection of hazardous chemicals produced at the Rocky Mountain Arsenal defense plant. The earthquakes continued after injection ceased in February 1966.[13] The disposal well was drilled through the flat-lying sedimentary rocks into the underlying older crystalline rocks more than 12,000 feet deep, and injection rates varied from 2 million gallons per month to as much as 5.5 million gallons per month.[14] Earthquake activity declined after 1967, but continued for the next two decades. Scientists concluded that the injection triggered the earthquakes, and that even after injection ceased, the migration of the underground pressure front continued for years and initiated earthquakes along an ancient fault system many miles away from the injection well.[15] As discussed below, the Rocky Mountain Arsenal earthquakes had many similarities to the recent increased earthquake activity in some deep-well injection activities of the United States, including, for example, injection near or in underlying crystalline bedrock, activation of fault systems miles away from the well, and migration of the pressure front away from the point of injection months or years after injection stopped.

Deep-Well Injection of Oil and Natural Gas Wastewaters

The number of earthquakes of M >3.0 in the central and eastern United States has increased dramatically since about 2009, from an average of approximately 20 per year between 1970 and 2000 to over 100 per year in the period 2010-2013.[16] *Figure 2* shows this increase in earthquake frequency as a steep increase in slope of the line of cumulative number of earthquakes starting in about 2004 but increasing sharply from about 2009, and departing from the relatively unchanging slope of the average number of earthquakes from 1970 to 2000, depicted as a dashed line.

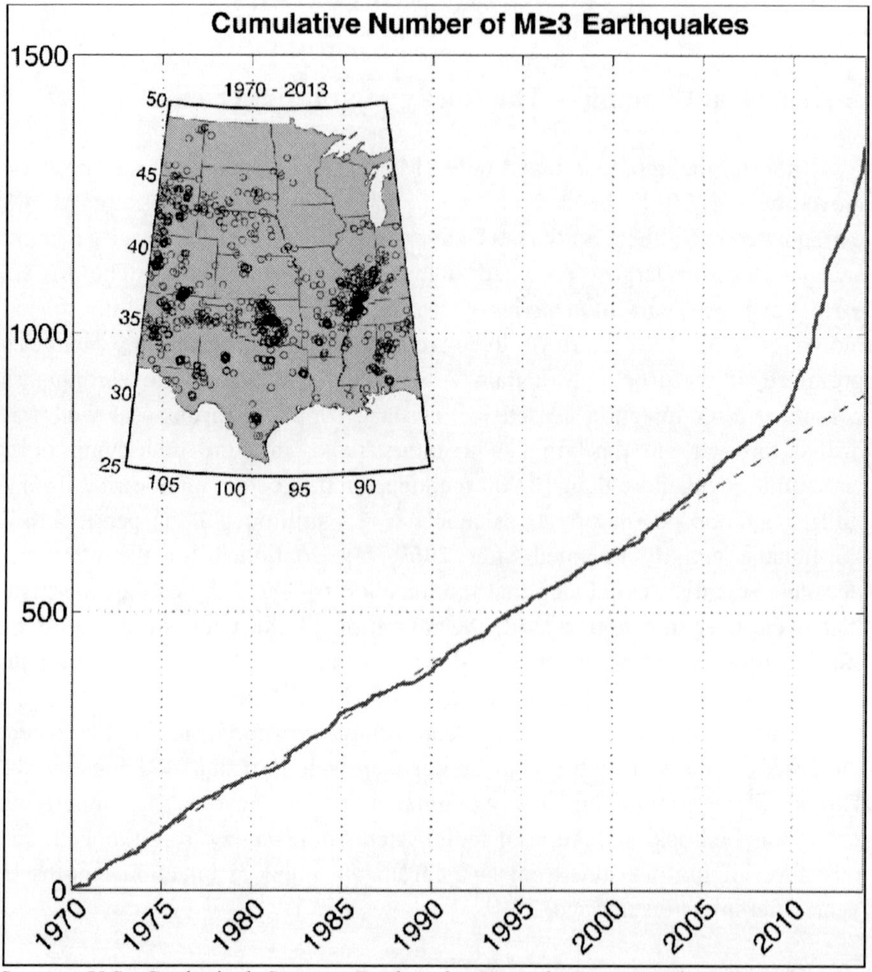

Source: U.S. Geological Survey, Earthquake Hazards Program, http:// earthquake.usgs.gov/research/induced/.

Notes: The dashed line corresponds to the long-term rate of about 20 earthquakes of M 3.0 or greater per year. A significant increase in the rate of these >M 3.0 earthquakes started around 2009.

Figure 2. Cumulative Number of Magnitude 3.0 or Greater Earthquakes in the Central and Eastern United States, 1970-2013.

States experiencing higher levels of seismic activity compared to the pre-2005 average include Arkansas, Colorado, Texas, New Mexico, Ohio, Oklahoma, and Virginia.[17] For some of these states, there is an increasing realization of a potential linkage between deep-well injection of oil and gas wastewaters and earthquakes, as the number of wells and volume of disposed wastewater have increased concomitant with increased domestic oil and gas production, particularly since about 2008 and 2009.[18] Several instances of suspected human-induced earthquakes that garnered media and national attention include:

- October 2008/May 2009—M 2.5-3.3 earthquakes near Dallas-Fort Worth, Texas;[19]
- August 2010/February 2011—earthquake swarm in central Arkansas, with M 4.7 earthquake on February 27, 2011, near Greenbrier, Arkansas;[20]
- August 2011—M 5.3 earthquake in the Raton Basin, northern New Mexico/southern Colorado;[21]
- December 2011—M 3.9 earthquake near Youngstown, OH;[22] and
- November 2011—M 5.6 earthquake near Prague, OK.[23]
- These examples are summarized below.

Colorado and New Mexico

An investigation of the seismicity in the Raton Basin of northern New Mexico and southern Colorado concluded that increased seismic activity since August 2001 was associated with deep-well injection of wastewater related to the production of natural gas from coal-bed methane fields.[24] The study linked the increased seismicity to two high-volume disposal wells that injected more than seven times as much fluid as the Rocky Mountain Arsenal well in the period leading up to an August 2011 M 5.3 earthquake in the Raton Basin.

Arkansas

A study of a 2010-2011 earthquake swarm in central Arkansas noted that the study area experienced an increase in the number of M 2.5 or greater earthquakes since 2009, when the first of eight deep-well injection disposal wells became operational.[25] The rate of M >2.5 earthquakes increased from 1 in 2007 to 2 in 2008, 10 in 2009, 54 in 2010, and 157 in 2011, culminating in a M 4.7 earthquake on February 27, 2011.[26] Although the area has a history of seismic activity, including earthquake swarms in the early 1980s, the study noted that 98% of the earthquakes during the 2010-2011 swarm occurred within 6 kilometers of one of the waste disposal wells. In response, the Arkansas Oil and Gas Commission (AOGC) imposed a moratorium on oil and gas wastewater disposal wells in a 1,150 square-mile area of central Arkansas. Four disposal wells were shut down following injection of wastewater from the Fayetteville Shale.

Texas

A study of increased seismicity near Dallas-Fort Worth and Cleburne, Texas, identified a possible linkage between high injection rates of oilfield-related wastewater and earthquakes of M 1.5 or greater, and found that all 24 of the most reliably located earthquake epicenters occurred within about 1.5 miles of one or more injection wells.[27] The study examined earthquakes occurring between 2009 and 2011, and noted that it was possible that some of the earthquakes had a natural origin, but that it was implausible that all were naturally occurring. The investigation showed a probable linkage between earthquakes and some high-volume injection wells, but also pointed out that in other regions of the study area there exist similar high-volume injection wells but no increased seismic activity. The study hypothesized that injection might only trigger earthquakes if the injected fluids reach suitably oriented nearby faults under regional tectonic stress.

Ohio

A study reported that the Youngstown, Ohio, area, where there were no known past earthquakes, experienced over 100 small earthquakes between January 2011 and February 2012.[28] The largest among the six felt earthquakes was an M 3.9 event that occurred on December 31, 2011. The study concluded that the earthquakes, which occurred within the Precambrian crystalline rocks lying beneath sedimentary rocks, were induced by fluid injection from a deep injection well. The study noted that the level of seismicity dropped after periods when the injection volumes and pressures were at their lowest levels, indicating that the earthquakes may have been caused by pressure buildup and then stopped when the pressure dropped.

Oklahoma

According to the U.S. Geological Survey and the Oklahoma Geological Survey, the rate of earthquakes in central Oklahoma increased by 50% between October 2013 and May 2014.[29] The number of M 3.0 or greater earthquakes was 145 in the first four months of 2014, which exceeded the previous record of 109 M 3.0 or greater earthquakes annually set in 2013, and which continues the trend of increasing seismic activity since about 2009 (*Figure 3*).[30]

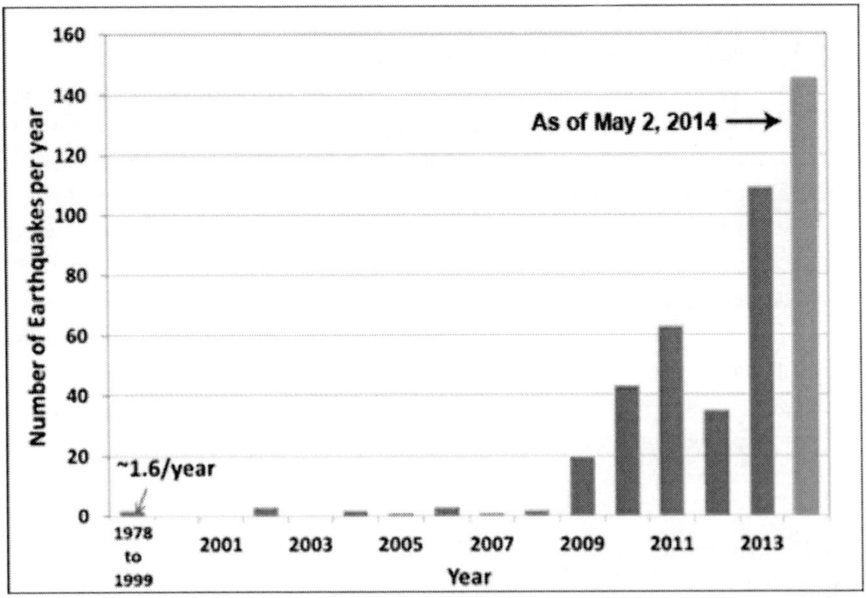

Source: U.S. Geological Survey, "Record Number of Oklahoma Tremors Raises Possibility of Damaging Earthquakes," Updated USGS-Oklahoma Geological Survey Joint Statement on Oklahoma Earthquakes, May 2, 2014, http:// earthquake.usgs.gov/contactus/golden/newsrelease_05022014.php. Modified by CRS.

Notes: Figure shows that 145 earthquakes of M 3.0 or greater occurred between January 1, 2014, and May 2, 2014. From 1978 through 2008, the state averaged about two M 3.0 or greater earthquakes per year.

Figure 3. Oklahoma Earthquakes of M 3.0 or Greater.

Since 2009, 20 earthquakes of M 4.0 to M 4.8 have struck central Oklahoma. The largest earthquake in Oklahoma history—M 5.6—occurred on November 5, 2011, near Prague, causing damage to several structures nearby. Central and northern Oklahoma are seismically active regions; however, the recent earthquake swarm does not seem to be due to typical, random changes in the rate of seismicity, according to a USGS statistical analysis.[31] The statistical analysis suggested that the increased rate of seismicity could be due to deep-well wastewater injection, and that an M 5.0 foreshock that preceded the M 5.6 earthquake on November 5, 2011, may have been induced by deep-well injection.[32] The M 5.0 event could then have triggered the subsequent M 5.6 event less than a day later.

The examples above indicate an increasing likelihood that in some instances, deep-well injection is linked to earthquakes, some greater than M 5.0. A human-induced M 6.0 or greater earthquake due to deep-well injection has not been observed, although scientists cannot rule out the possibility that one could occur in the future. However, the great majority of deep injection wells in the United States (UIC Class II) appear to be aseismic for earthquakes of M 3.0 or more.[33] Some observers conclude that most wells permitted for deep-well injection are in geologic formations that likely have a low risk of failure leading to damaging earthquakes, if the injected fluids remain in the intended geologic structure.[34] The largest earthquakes apparently triggered by deep-well injection involved faulting that was deeper than the injection interval, suggesting to some that transmitting pressure from the injection point to deeper zones in basement rocks— below the sedimentary layers—increases the potential for triggering earthquakes.[35]

Hydraulic Fracturing

Hydraulic fracturing (often referred to as fracking) is the process of injecting a slurry of water, chemicals, and sand at high pressure to fracture oil- and gas-bearing rocks in order to provide permeable pathways to extract hydrocarbons.[36] Fracking has been employed with increasing frequency over the past decade or so to produce oil and natural gas from "unconventional" formations (e.g., shale)—those geologic strata that contained hydrocarbons but because of natural impermeability were not exploitable by conventional oil and gas producing methods. Fracking intentionally propagates fractures in the rocks to improve permeability. Fracking induces microseismicity, mostly less than M 1.0, too small to feel or cause damage. In some cases, fracking has led to earthquakes larger than M 2.0, including at sites in Oklahoma, Ohio, England, and Canada. Hydraulic fracturing is generally thought to present less of a risk than disposal wells for inducing large earthquakes, because the injections are short-term and add smaller amounts of fluid into the subsurface compared to most disposal wells.

Canada

Between April 2009 and July 2011, and over a five-day period in December 2011, nearly 40 seismic events were recorded in the Horn River Basin, northeast British Columbia, ranging from M 2.2 to M 3.8.[37] A subsequent investigation indicated that the seismic events were linked to fluid injection during hydraulic fracturing activities near pre-existing faults. In contrast to the vast majority of hydraulic fracturing injection activities, which cause earthquakes not felt at the surface (e.g., over 8,000 fracking completions in the Horn River Basin without any associated anomalous seismicity), these anomalous seismic events were felt at the ground surface.

England

In Blackpool, England, hydraulic fracturing injection activities led to a series of small earthquakes ranging up to M 2.3, between March 28, 2011, and May 28, 2011.[38] These seismic events were not large enough to be felt at the surface, but were strong enough to deform some of the well casing on the horizontal portion of the production well used for fracking the shale gas-bearing formation.

Oklahoma

In south-central Oklahoma, hydraulic fracturing injections between January 16, 2011, and January 22, 2011, induced a series of 116 earthquakes of M 0.6 to M 2.9, according to one study.[39] The study concluded that the lack of similar seismic activity prior to the fracking, and after fracking ceased, among other factors, linked the fracking activities to the earthquakes. More recently presented work on the link between hydraulic fracturing and earthquakes in Oklahoma seems to further strengthen the association between fracking and earthquakes that may rarely exceed M 3.0 or even M 4.0 in some cases.[40] The more recent work in Oklahoma also indicated that the vast majority of fracking operations did not create anomalous seismicity.

Ohio

Recently published research on a series of small earthquakes in Harrison County, Ohio, indicated that hydraulic fracturing operations affected a previously unmapped fault in the Precambrian crystalline rocks lying below the sedimentary rocks that were being hydraulically fractured.[41] None of the Harrison County earthquakes exceeded magnitude 2.2, but various lines of evidence suggested that the fault responsible for the small earthquake was triggered by hydraulic fracturing operations. Some seismic activity possibly related to fracking in the Marcellus Shale and the underlying Utica Shale led to changes in how Ohio permits wells.[42] The permitting changes include requirements to install seismic monitoring equipment if drilling will take place within 3 miles of a known fault, or in an area with seismic activity greater than M 2.0. Further, if the monitors detect a seismic event greater than M 1.0, activities at the site must cease while the cause is investigated.

Other Issues

One of the major shale gas plays in the United States, the Marcellus Shale, which underlies western Pennsylvania, and portions of New York, West Virginia, and Ohio, occurs in a region of relatively low levels of natural seismic activity. Despite thousands of hydraulic fracturing operations in the past decade or so, only a handful of M 2.0 or greater earthquakes were detected within the footprint of the Marcellus Shale, as measured by a regional seismographic network.[43] The earthquake activity recorded in the Youngstown, OH, region was related to deep-well injection of waste fluids from the development of Marcellus Shale gas, but was not associated with hydraulic fracturing of Marcellus Shale in Pennsylvania.[44]

The linkage between hydraulic fracturing itself and the potential for generating earthquakes large enough to be felt at the ground surface is an area of active research. It appears to be the case that hydraulic fracturing operations mostly create microseismic activity—too small to be felt— associated with fracturing the target formation to release trapped natural gas or oil. However, if the hydraulic fracturing fluid injection affects a nearby fault, there exists the potential for larger earthquakes possibly strong enough to be felt at the surface, as was the case in the Horn River Basin of western Canada.

Overview of the Current Regulatory Structure Regarding Induced Seismicity

The National Research Council (NRC) estimates that conventional oil and gas production and hydraulic fracturing, combined, generate more than 800 billion gallons of fluid each year. More than one-third of this volume is injected for permanent disposal in Class II injection wells.[45] Deep-well injection has long been the environmentally preferred option for managing produced brine and other wastewater associated with oil and gas production poses. However, the development of unconventional formations using high-volume hydraulic fracturing has contributed significantly to a growing volume of wastewater requiring disposal. Recent incidents of seismicity in the vicinity of disposal wells have drawn renewed attention to laws, regulations, and policies governing wastewater management and have generated various responses at the federal and state levels. This section of the report reviews the current regulatory framework for managing underground injection and identifies several federal and state initiatives in response to concerns surrounding Class II disposal and induced seismicity.

EPA Regulation of Underground Injection Activities

The principal law authorizing federal regulation of underground injection activities is the Safe Drinking Water Act (SDWA) of 1974, as amended.[46] The law specifically directs EPA to promulgate regulations for state underground injection control (UIC) programs to prevent underground injection that endangers drinking water sources.[47] Historically, EPA has not regulated oil and gas production wells, and as amended in 2005, the SDWA explicitly excludes the regulation of underground injection of fluids or propping agents (other than diesel fuels) associated with hydraulic fracturing operations related to oil, gas, and geothermal production activities.[48]

The SDWA authorizes states to assume primary enforcement authority (primacy) for the UIC program for any or all classes of injection wells. EPA must delegate this authority, provided that the state program meets certain statutory and EPA requirements. If a state's UIC program plan is not approved, or if a state chooses not to assume program responsibility, then EPA implements the UIC program in that state.

For oil-and-gas-related injection operations (such as produced water disposal through Class II wells), the law allows states to administer the UIC program using state rules rather than following EPA regulatory requirements, provided a state demonstrates that it has an effective program that prevents underground injection that endangers drinking water sources.[49] Most oil and gas states have assumed primacy for Class II wells under this provision.

Under the UIC program, EPA, states, and tribes regulate more than 800,000 injection wells. To implement the UIC program as mandated by the SDWA, EPA has established six classes of underground injection wells based on categories of materials injected by each class. In addition to the similarity of fluids injected, each class shares similar construction, injection depth, design, and operating techniques. The wells within a class are required to meet a set of appropriate performance criteria for protecting underground sources of drinking water (USDWs).[50] *Figure 4* provides an illustration of the six well classes established by EPA to implement the UIC program.

Source: U.S. Environmental Protection Agency, Underground Injection Control, Typical Injection Wells.

Figure 4. Federally Regulated Underground Injection Wells.

Class II includes wells used to inject fluids associated with oil and gas production. Class II wells may be used for three broad purposes: (1) to dispose of brines (salt water) and other fluids associated with oil and gas production; (2) to store petroleum natural gas; or (3) to inject fluids to enhance recovery of oil and gas from conventional fields. There are roughly 180,500 Class II wells across the United States. Based on historical averages, roughly 80% of the Class II wells are enhanced recovery wells, and 20% are disposal wells (Class IId wells).[51]

Table 1 provides descriptions of the injection well classes and subcategories and estimated numbers of wells.

Table 1. UIC Program: Classes of Injection Wells and Nationwide Numbers

Well Class	Purpose and Uses
Class I	Wells inject hazardous wastes, industrial non-hazardous liquids, or municipal wastewater beneath the lowermost underground source of drinking water (USDW). (680 wells, including 117 hazardous waste wells)
Class II	Wells inject brines and other fluids associated with oil and gas production, and hydrocarbons for storage. The wells inject fluids beneath the lowermost USDW. (>180,000 wells) Types of Class II wells:[a] • Enhanced Recovery (ER) Wells: Separate from, but often surrounded by, production wells, these wells are used to inject produced water (brine), fresh water, steam, polymers, or carbon dioxide (CO2) into oil-bearing formations to recover additional oil (and sometimes gas) from production wells. These wells also may be used to maintain reservoir pressure. Approximately 80% of Class II wells are ER wells. • Disposal wells: Produced water and other fluids associated with oil and gas production (including flowback from hydraulic fracturing operations) are injected into these wells for permanent disposal. Approximately 20% of Class II wells are disposal (Class IId) wells. • • Hydrocarbon storage wells: More than 100 Class II wells are used to inject hydrocarbons (petroleum and natural gas) into underground formations for storage.
Class III	Class III wells inject fluids associated with solution mining of minerals (e.g., salt and uranium) beneath the lowermost USDW. (22,131 wells)
Class IV	Class IV wells inject hazardous or radioactive wastes into or above USDWs. These wells are banned unless authorized under a federal or state groundwater remediation project. (33 wells)
Class V	Class V includes all injection wells not included in Classes I-IV, including experimental wells. Class V wells often inject non-hazardous fluids into or above USDWs and are typically shallow, on-site disposal systems (e.g., cesspools and stormwater drainage wells). Some deep Class V wells (e.g., geothermal energy and aquifer storage wells) inject below USDWs. (>450,000 wells)
Class VI	Class VI, established in 2010, includes wells used for the geologic sequestration of carbon dioxide (CO2). (2 permits approved in 2014)

Source: U.S. Environmental Protection Agency, Underground Injection Control Program, Classes of Wells, and Class II Wells—Oil and Gas Related Injection Wells (Class II), http://water.epa.gov/type/groundwater/uic/wells.cfm, and UIC well surveys.

Notes: Regulations for Class I (hazardous waste) and Class VI (CO2 sequestration) wells include evaluation of seismic risk among requirements to prevent movement of fluids out of the injection zone to protect USDWs.

[a] A Class II permit would be required for an oil, gas, or geothermal production well if diesel fuels were to be used in the hydraulic fracturing fluid.

Consideration of Seismicity in EPA UIC Regulations

The SDWA does not mention seismicity; rather, the law's UIC provisions authorize EPA to regulate underground injection to prevent endangerment of underground sources of drinking water. However, seismicity has the potential to affect drinking water quality through various means (e.g., by damaging the integrity of a well, or creating new fractures and pathways for fluids to reach groundwater). EPA UIC regulations include various requirements aimed at protecting USDWs by ensuring that injected fluids remain in a permitted injection zone. Some of these measures also could reduce the likelihood of triggering seismic events. For example, injection pressures for Class II (and other) wells may not exceed a pressure that would initiate or propagate fractures in the confining zone adjacent to a USDW.[52] As a secondary benefit, limiting injection pressure can prevent fractures that could act as conduits through which injected fluids could reach an existing fault.

EPA regulations for two categories of injection wells—Class I hazardous waste disposal wells, and Class VI wells for geologic sequestration of CO_2—specifically address evaluation of seismicity risks with siting and testing requirements. For Class I wells, EPA regulations include minimum criteria for siting hazardous waste injection wells, requiring that wells must be limited to areas that are geologically suitable. The UIC Director (i.e., the delegated state or EPA) is required to determine geologic suitability based upon an "analysis of the structural and stratigraphic geology, the hydrogeology, and the seismicity of the region."[53] Testing and monitoring requirements for Class I wells state that "the Director may require seismicity monitoring when he has reason to believe that the injection activity may have the capacity to cause seismic disturbances."[54]

For Class VI CO_2 sequestration wells, EPA regulations similarly require evaluation of seismicity risks through siting and testing requirements. In determining whether to grant a permit, the UIC Director must consider various factors, including potential for seismic activity.

> Prior to the issuance of a permit for the construction of a new Class VI well or the conversion of an existing Class I, Class II, or Class V well to a Class VI well, the owner or operator shall submit, ... and the Director shall consider ... information on the seismic history including the presence and depth of seismic sources and a determination that the seismicity would not interfere with containment.[55]

EPA regulations for oil and gas wastewater disposal wells (or other Class II wells) do not include these provisions, or otherwise address seismicity; however, the regulations give discretion to UIC Directors to include in individual permits additional requirements as needed to protect underground sources of drinking water.[56] Again, for the purpose of protecting drinking water sources, permits for all Class I, II, and III wells must contain specified operating conditions, including "a maximum operating pressure calculated to avoid initiating and/or propagating fractures that would allow fluid movement into a USDW."[57] Regulations for Class I wells further specify that "injection pressure must be limited such that no fracturing of the injection zone occurs during operation."[58]

Outside of regulations, EPA recently has taken steps to address induced seismicity concerns associated with Class II disposal wells. For example, EPA Region III now evaluates induced seismicity risk factors when considering permit applications for Class II wells. (Region III directly implements the UIC program in Pennsylvania and Virginia.)[59] In responding to public comments on a Class II well permit application, the regional office noted the following:

> Although EPA must consider appropriate geological data on the injection and confining zone when permitting Class II wells, the SDWA regulations for Class II wells do not require specific consideration of seismicity, unlike the SDWA regulations for Class I wells used for the injection of hazardous waste.... Nevertheless, EPA evaluated factors relevant to seismic activity such as the existence of any known faults and/or fractures and any history of, or potential for, seismic events in the areas of the Injection Well as discussed below and addressed more fully in "Region 3 framework for evaluating seismic potential associated with UIC Class II permits, updated September, 2013."[60]

EPA expects that seismic activity is likely to be induced by Class II well injections only when several conditions are present: "(1) there is a fault in a near-failure state of stress; (2) the fluid injected has a path of communication to the fault: and (3) the pressure exerted by the fluid is high enough and lasts long enough to cause movement along the fault line."[61]

Federal Initiatives to Address Induced Seismicity

As discussed above, the Safe Drinking Water Act does not directly address seismicity; rather, the law authorizes EPA to regulate subsurface injections to prevent endangerment of drinking water sources. In 2011, in response to earthquake events in Arkansas and Texas, EPA asked the Underground Injection Control National Technical Workgroup to "develop technical recommendations to inform and enhance strategies for avoiding significant seismicity events related to Class II disposal wells." The workgroup specifically was asked to address concerns that induced seismicity associated with Class II disposal wells could cause injected fluids to move outside the containment zone and endanger drinking water sources. (EPA has not initiated any rulemaking to address this issue.)

The UIC workgroup completed a draft report in late 2012.[62] EPA plans to release the final report in early 2015. The final report is expected to include practical tools and best practices to address injection-induced seismicity: it will not constitute formal agency guidance. EPA requested that the report contain the following specific elements:

- Comparison of parameters identified as most applicable to induced seismicity with the technical parameters collected under current regulations.
- Decision-making model/conceptual flow chart to:
- provide strategies for preventing or addressing significant induced seismicity,
- identify readily available applicable databases or other information,
- develop site characterization check list, and
- explore applicability of pressure transient testing and/or pressure monitoring techniques.
- Summary of lessons learned from case studies.
- Recommended measurement or monitoring techniques for higher risk areas.
- Applicability of conclusions to other well classes.
- Define specific areas of research as needed.[63]

The Department of Energy (DOE) conducts a research program to promote development of the nation's geothermal resources, including development of enhanced geothermal systems (EGS). The development of

EGS can enable uneconomic hydrothermal systems to produce geothermal energy on a large scale. However, the process of injecting fluids to enhance permeability of hydrothermal systems may trigger a seismic event. In 2012, DOE released an Induced Seismicity Protocol to mitigate risks associated with the development of these systems.[64] Some of the approaches and mitigation measures included in the DOE protocol may be applicable to issues posed by Class II disposal wells.

State Initiatives

Several states and state organizations have been assessing the possible relationship between injection wells and seismic activity. In 2013, the Ground Water Protection Council (GWPC)[65] published a white paper on assessing and managing risk of induced seismicity by underground injection.[66] In March 2014, the Interstate Oil and Gas Compact Commission (IOGCC)[67] and the GWPC formed an Induced Seismicity Work Group with state regulatory agencies and geological surveys to "proactively discuss the possible association between recent seismic events occurring in multiple states and injection wells."[68]

Additionally, several states have strengthened requirements for Class II disposal wells in response to recent seismic events that appear to be injection related. Policy and regulatory developments adopted or under consideration by several states are outlined briefly below. Typically, these states have expanded their standard permit application packages to include, for example, requirements for additional existing geologic information and studies, and stricter operating requirements. Also, some states have banned the drilling of injection wells in geologic zones of known seismic risk.

Arkansas

In response to the Guy-Greenbrier earthquake swarm associated with injections of wastewater from shale gas production, the Arkansas Oil and Gas Commission (AOGC) in 2010 imposed a moratorium on new disposal wells in the vicinity of the increased seismic activity, and required operators of seven existing wells in the area to provide bi-hourly injection rates and pressures.[69]

In 2011, the AOGC revised rules governing Class II wells and established a permanent moratorium zone in the area of a major fault system. The state banned new disposal wells and required plugging of four existing wells within the zone.[70] The rules also require Commission approval and a public hearing before any Class II wells within specified distances from the Moratorium Zone Deep Fault or a regional fault can be drilled, deepened, reentered, or recompleted. Class II wells proposed for disposal above or below the Fayetteville Shale formation are subject to new siting and spacing requirements, and permit applicants are required to provide to the state information on the structural geology of an area proposed for a new disposal well. For existing disposal wells, permit holders must install flow meters and submit injection volume and pressure information at least daily.[71]

Colorado

The Colorado Oil and Gas Conservation Commission (COGCC) has identified in existing rules and policies various requirements that reduce the likelihood of induced seismicity.[72] These safeguards, which are imposed through the permitting process, include setting limits on injection volume and rate, and requiring that the maximum allowable injection pressure is set below the fracturing pressure for the injection zone.[73] In 2011, COGCC expanded the UIC permit review process specifically to minimize risk of induced seismicity from oil and gas wastewater disposal. The changes followed a significant earthquake near wells injecting wastewater produced from a coalbed methane field. The COGCC now has the Colorado State Geologist (CGS) review permit applications to evaluate the area for the proposed well site for seismic activity. The CGS reviews state geologic maps, the USGS earthquake database, and area-specific information. After reviewing the geologic history and maps of the area for faults, the CGS may recommend a more detailed review of subsurface geology or seismic monitoring prior to new drilling. Additionally, the Division of Water Resources conducts a review of the proposed injection zone.[74]

In July 2014, the COGGC reported that the commission is working with the Colorado Geological Survey, USGS researchers, and state universities to establish an induced seismicity advisory group. Issues for consideration by the advisory group include development of a more comprehensive statewide seismicity monitoring network and improved guidance for managing high volume injection.

Ohio

Following the Youngstown earthquakes in 2011 associated with Class II disposal wells, the Ohio Department of Natural Resources (ODNR) prohibited all drilling into the Precambrian basement rock and added new permit requirements for Class II disposal wells to improve site assessment and collection of more comprehensive information. The rules became effective in October 2012, and are implemented on a well-by-well basis through the permitting process. The supplemental permit application requirements could include pressure fall-off testing, geological evaluation of potential faulting, seismic monitoring program (baseline and active injection), minimum geophysical logging suite, radioactive tracer or spinner survey, and any other tests deemed necessary by the Division of Oil and Gas Resources Management.[75] Before approving a new Class II disposal well, state officials now review existing geologic data for known faulted areas. ODNR will also require companies to run a complete suite of geophysical logs on newly drilled Class II disposal wells. Companies are required to give ODNR a copy of the log suite and where required, provide analytical interpretation of the logging. For all new Class II permit applications, ODNR requires installation of monitoring technologies, including a continuous pressure monitoring system and an automatic shutoff system.[76]

In 2014, ODNR drafted new rules for construction of horizontal production wells that are to be hydraulically fractured (i.e., shale oil and gas wells) in response to seismic activity the state determined had a "probable connection to hydraulic fracturing near a previously unknown microfault."[77] The draft rules include standards for design, approval, and construction of horizontal well sites, and would strengthen drilling permit conditions for wells located near faults or areas linked to previous seismic activity.[78]

Texas

In November 2014, the Texas Railroad Commission (RRC) published amendments to the state's oil and gas rules to incorporate requirements related to seismic events in connection with wastewater disposal permits, monitoring, and reporting.[79] Several of the new requirements are listed below.[80]

- Applicants for disposal well permits are required to provide information from the USGS regarding the locations of any historical seismic events within 100 square miles of the proposed well site.
- A permit for a Class II disposal well "may be modified, suspended, or terminated if injection is likely to be or determined to be contributing to seismic activity."[81]
- The RRC may require permit applicants to provide additional information (e.g., logs, geologic cross-sections, and pressure front boundary calculations) if the well is to be located in an area where conditions may increase the risk that fluids will not be confined in the

injection interval. (Such conditions may include complex geology, proximity of the basement rock to the injection interval, transmissive faults, and/or a history of seismic events using available USGS information.)

- Operators may be required to conduct more frequent monitoring and reporting of disposal well injection pressures and rates if certain conditions are present that could increase the risk that fluids will not be confined to the injection interval.

Although states have taken various actions in response to recent seismic events and wastewater injection, additional regulatory actions could result from the IOGCC and GWPC Induced Seismicity Work Group as state regulatory agencies and geological surveys continue to evaluate this issue.

Conclusion

The scientific understanding of linkages between deep-well injection of waste fluids from oil and gas production, and from hydraulic fracturing operations, is rapidly evolving. This poses a challenge to state and federal policy makers who are tasked with making policy, regulatory, and permitting decisions in a relatively short time frame, concomitant with the evolving scientific study and understanding, and given public concern over the possibility of damaging earthquakes from some of the deep disposal wells. Some states have already implemented changes to their regulatory and permitting requirements, as discussed above. The vast majority of deep-well injection wells and hydraulic fracturing wells do not appear to be associated with significant seismic events. Additional geologic studies and reviews adopted by some states should address some potential risks; however, it is likely that states and possibly the federal government will continue to explore ways to understand and mitigate against the possibility of damaging earthquakes caused by a small number of wells.

In early 2015, EPA plans to publish a report outlining best practices to address seismic events associated with oil and gas wastewater injection. Congress may be interested in oversight of EPA's UIC program and, more broadly, in federally sponsored research on the relationship between energy development activities and induced seismicity.

Only a small fraction of the more than 30,000 U.S. wastewater disposal wells appears to be problematic for causing damaging earthquakes. However, such incidents may raise questions as to whether other energy-related activity—specifically, underground injection for carbon dioxide sequestration—may present similar risks.

End Notes

[1] U.S. Geological Survey/Oklahoma Geological Survey joint statement, "Record Number of Oklahoma Tremors Raises Possibility of Damaging Earthquakes," May 2, 2014, http://earthquake.usgs.gov/regional/ceus/products/ newsrelease_05022014.php.

[2] Ibid.

[3] EPA has established regulations for six classes of injection wells, including Class II wells used for the injection of fluids for enhanced oil and gas recovery and wastewater disposal. See section on "EPA Regulation of Underground Injection Activities" for more information.

[4] William L. Ellsworth, "Injection-Induced Earthquakes," Science, vol. 341, July 12, 2013, http://www.sciencemag.org/ content/341/6142/1225942.full. Hereinafter Ellsworth, 2013.

[5] National Research Council, "Induced Seismicity Potential in Energy Technologies," 2013, p. vii. Hereinafter referred to as NRC, 2013.

[6] Ellsworth, 2013.

[7] Ibid.

[8] According to the National Research Council report, seismic events likely related to energy development have been documented in Alabama, Arkansas, California, Colorado, Illinois, Louisiana, Mississippi, Nebraska, Nevada, Ohio, Oklahoma, and Texas. NRC, 2013, p. 6.

[9] Ibid.

[10] William Leith, Senior Science Advisor for Earthquakes and Geologic Hazards, U.S. Geological Survey, "USGS Research into the Causes & Consequences of Injection-Induced Seismicity," presentation at the U.S. Energy Association, Oct. 30, 2014, http://www. usea.org/sites/default/files/event-/Leith%20induced%20for%20DOEUSEA% 20Oct14.pdf.

[11] For a more general discussion of earthquakes, see CRS Report RL33861, Earthquakes: Risk, Detection, Warning, and Research, by Peter Folger.

[12] U.S. Geological Survey FAQs, at http://earthquake.usgs.gov/learn/faq/; and Magnitude/ Intensity Comparison, at http://earthquake.usgs.gov/learn/topics/ mag_vs_ int.php.

[13] J. H. Healy et al., "The Denver Earthquakes," Science, vol. 161, no. 3848 (September 27, 1968), pp. 1301-1310.

[14] Healy et al., 1968.

[15] Ellsworth, 2013.

[16] U.S. Geological Survey, "Man-Made Earthquakes Update," January 17, 2014, http://www. usgs.gov/blogs/features/ usgs_top_story/man-made-earthquakes/.

[17] Ellsworth, 2013.

[18] Ibid.

[19] Cliff Frohlich et al., "Dallas-Fort Worth Earthquakes Coincident with Activity Associated with Natural Gas Production," The Leading Edge, vol. 29, no. 3 (2010), pp. 270-275.

[20] U.S. Geological Survey, Earthquake Hazards Program, "Poster of the 2010-2011 Arkansas Earthquake Swarm," http://earthquake.usgs.gov/earthquakes/eqarchives /poster/2011/ 20110228.php.

[21] J. L. Rubinstein, W. L. Ellsworth, and A. McGarr, "The 2001-Present Triggered Seismicity Sequence in the Raton Basin of Southern Colorado/Northern New Mexico," talk delivered at the Seismological Society of America Annual Meeting, Salt Lake City, UT, April 19, 2013, pp. Abstract #13-206.

[22] Won-Young Kim, "Induced Seismicity Associated With Fluid Injection Into a Deep Well in Youngstown, Ohio," Journal of Geophysical Research—Solid Earth, vol. 118, no. 7 (July 19, 2013), pp. 3506-3518.

[23] Danielle F. Sumy et al., "Observations of Static Coulomb Stress Triggering of the November 2011 M 5.7 Oklahoma Earthquake Sequence," Journal of Geophysical Research—Solid Earth, vol. 119, no. 3 (March 2014), http://onlinelibrary.wiley.com/ doi/10.1002 /2013JB010612/abstract.

[24] Rubinstein et al., 2013.

[25] S. Horton, "Disposal of Hydrofracking Waste Fluid by Injection Into Subsurface Aquifers Triggers Earthquake Swarm in Central Arkansas with Potential for Damaging Earthquake," Seismological Research Letters, vol. 83, no. 2 (2012), pp. 250-260.

[26] U.S. Geological Survey, Earthquake Hazards Program, "Poster of the 2010-2011 Arkansas Earthquake Swarm."

[27] Cliff Frohlich, "Two-Year Survey Comparing Earthquake Activity and Injection-Well Locations in the Barnett Shale, Texas," Proceedings of the National Academy of Sciences, vol. 109, no. 35 (August 28, 2012), pp. 13934-13938.

[28] Won-Young Kim, "Induced Seismicity Associated With Fluid Injection Into a Deep Well in Youngstown, Ohio," Journal of Geophysical Research—Solid Earth, vol. 118, no. 7 (July 19, 2013), pp. 3506-3518.

[29] U.S. Geological Survey, "Record Number of Oklahoma Tremors Raises Possibility of Damaging Earthquakes," Joint USGS-Oklahoma Geological Survey joint statement, May 2, 2014, http://earthquake.usgs.gov/regional/ceus/products/ newsrelease_05022014.php.

[30] Ibid.

[31] Ibid.

[32] Danielle F. Sumy et al., "Observations of Static Coulomb Stress Triggering of the November 2011 M 5.7 Oklahoma Earthquake Sequence," Journal of Geophysical Research—Solid Earth, vol. 119, no. 3 (March 2014).

[33] Ellsworth, 2013.

[34] Ibid.

[35] Ellsworth, 2013.

[36] This process has also been used for enhanced geothermal energy development, in which rocks are fractured to create permeable pathways to circulate fluids at depth. The fluids are heated by the Earth's natural heat, and then recirculated to the surface to drive a turbine and generate electricity.

[37] BC Oil and Gas Commission, Investigation of Observed Seismicity in the Horn River Basin, August 2012, http://www.bcogc.ca/node/8046/download.

[38] Christopher A. Green, Peter Styles, and Brian J. Baptie, Preese Hall Shale Gas Fracturing, Review & Recommendations for Induced Seismic Mitigation, April 2012, https:// www.gov.uk/government/uploads/system/ uploads/attachment_data/file/15745/5075-preese-hall-shale-gas-fracturing-review.pdf.

[39] Austin Holland, "Earthquakes Triggered by Hydraulic Fracturing in South-Central Oklahoma," Bulletin of the Seismological Society of America, vol. 103, no. 3 (June 2013), pp. 1784-1792.

[40] Austin Holland, "Induced Seismicity 'Unknown Knowns': the Role of Stress and Other Difficult to Measure Parameters of the Subsurface," Presentation at the U.S. Energy Association Symposium: Subsurface Technology and Engineering Challenges and R&D Opportunities, Washington, DC, October 30, 2014, http://www.usea.org/event/ subsurface-technology-engineering-challenges-and-rd-opportunities-stress-state-and-induced.

[41] Paul A. Friberg, Glenda M. Besana-Ostman, and Ilya Dricker, "Characterization of an Earthquake Sequence Triggered by Hydraulic Fracturing in Harrison County, Ohio," Seismological Research Letters, vol. 85, no. 6 (November/December 2014), pp. 1-13.

[42] Ohio Department of Natural Resources, Ohio Announces Tougher Permit Conditions for Drilling Activities Near Faults and Areas of Seismic Activity, April 11, 2014, http://ohiodnr.gov/news/post/ohio-announces-tougher-permitconditions-for-drilling-activities-near-faults-and-areas-of-seismic-activity.

[43] Ellsworth, 2013.

[44] Ibid.

[45] National Research Council, Induced Seismicity Potential in Energy Technologies, Committee on Induced Seismicity Potential in Energy technologies, National Academy Press, Washington, DC, 2012, p 110.

[46] The Safe Drinking Water Act of 1974 (P.L. 93-523) authorized the UIC program at EPA. UIC provisions are contained in SDWA Part C, §§1421-1426; 42 U.S.C. §§300h-300h-5.

[47] 42 U.S.C. §300h(d). SDWA Section 1421.

[48] The Energy Policy Act of 2005 (EPAct 2005; P.L. 109-58, §322) amended the definition of "underground injection," SDWA§1421(d), to expressly exempt hydraulically fractured oil, gas, or geothermal production wells from the UIC program unless diesel fuels are used in the fracturing fluid.

[49] SDWA Section 1425 requires a state to demonstrate that its UIC program meets the requirements of Section 1421(b)(1)(A) through (D) and represents an effective program (including adequate record keeping and reporting) to prevent underground injection that endangers underground sources of drinking water. To receive approval under Section 1425's optional demonstration provisions, a state program must include permitting, inspection, monitoring, and record-keeping and reporting requirements.

[50] EPA regulations define a USDW to mean an aquifer or part of an aquifer that (a) supplies a public water system, or contains a sufficient quantity of groundwater to supply a public water system, and currently supplies drinking water for human consumption, or contains fewer than 10,000 milligrams per liter (mg/L or parts per million) total dissolved solids; and (b) is not an "exempted aquifer." 40 C.F.R. 144.3.

[51] U.S. Environmental Protection Agency, Class II Wells—Oil and Gas Related Injection Wells (Class II), http://water.epa.gov/type/groundwater/uic/class2/index.cfm, May 9, 2012.

[52] 40 C.F.R. §146.23(a)(1).

[53] 40 C.F.R. §146.62(b)(1).

[54] 40 C.F.R. §146.68(f).

[55] 40 C.F.R. §146.82(a)(3)(v).

[56] Relevant provisions for Class II wells are published at 40 C.F.R. §144.12(b) and 40 C.F.R. §144.52(a)(9).

[57] U.S. Environmental Protection Agency, Technical Program Overview: Underground Injection Control Regulations, EPA 816-R-02-005, Revised July 2001, p. 65, http://water.epa.gov /type/groundwater/uic/upload/ 2004_5_3_uicv_techguide_uic_tech_overview_uic_regs.pdf.

[58] Ibid., p. 66.

[59] EPA also directly implements the UIC program for other oil and gas producing states, including Kentucky, Michigan, and New York.

[60] U.S. Environmental Protection Agency Region III, Response to Comments for the Issuance of an Underground Injection Control (UIC) Permit for Windfall Oil and Gas, Inc., 2013, pp. 3-9, http://www.epa.gov/reg3wapd/pdf/ public_notices/WindfallResponsivenessSummary.pdf.

[61] Ibid., p. 4.

[62] U.S. Environmental Protection Agency, Minimizing and Managing Potential Impacts of Induced-Seismicity from Class II Disposal Wells: Practical Approaches, draft report of the Underground Injection Control National Technical Workgroup, November 27, 2012.

[63] Ibid, p. A-1-2. Technical Recommendations to Address the Risk of Class II Disposal Induced Seismicity, Office of Ground Water and Drinking Water, UIC National Technical Workgroup Project Topic #2011-3, July 2011.

[64] Emie Majer, James Nelson, and Ann Roberson-Tait, et al., Protocol for Addressing Induced Seismicity Associated with Enhanced Geothermal Systems, U.S. Department of Energy, Office of Energy Efficiency and Renewable Energy, DOE/EE-0662, January 2012, 45 pp., https://www1.eere.energy.gov/geothermal/pdfs/ geothermal_seismicity_protocol_012012.pdf.

[65] The Ground Water Protection Council (GWPC) represents state groundwater protection and underground injection control agencies, http://www.gwpc.org/.

[66] Ground Water Protection Council, White Paper Summarizing a Special Session on Induced Seismicity, February 2013, http://www.gwpc.org/sites/default/files /events/white%20paper %20-%20final_0.pdf.

[67] The Interstate Oil and Gas Compact Commission is a multi-state agency that "serves as the collective voice of member governors on oil and gas issues and advocates states' rights to govern petroleum resources within their borders." The commission works with other stakeholders and is chartered to "efficiently maximize oil and natural gas resources through sound regulatory practices while protecting health, safety and the environment." http://iogcc.publishpath.com/.

[68] States First Initiative, *States Team Up to Assess Risk of Induced Seismicity*, April 29, 2014, http://www.statesfirstinitiative.org.

[69] U.S. Environmental Protection Agency, *Minimizing and Managing Potential Impacts of Induced-Seismicity from Class II Disposal Wells: Practical Approaches*, draft report of the Underground Injection Control National Technical Workgroup, November 27, 2012, p. 15.

[70] Specifically, the rules state: "Unless otherwise approved by the Commission after notice and a hearing, no permit to drill, deepen, re-enter, recomplete or operate a Class II Disposal or Class II Commercial Disposal Well may be granted for any Class II or Class II Commercial Disposal wells in any formation within [a prescribed] area ("Moratorium Zone"). AOGC Rule H-1, Section (s)(2)."

[71] Arkansas Oil and Gas Commission: General Rule H – Class II Wells, Rule H-1: Class II Disposal and Class II Commercial Disposal Well Permit Application Procedures, Section (s).

[72] The Colorado Oil and Gas Conservation Commission (COGCC) administers the UIC program in accordance with EPA regulations. 40 C.F.R. §§144-147.

[73] Colorado Oil and Gas Conservation Commission, *COGCC Underground Injection Control and Seismicity in Colorado*, January 19, 2011.

[74] Colorado Department of Natural Resources, "COGCC Underground Injection Control and Seismicity in Colorado," Colorado Oil and Gas Conservation Commission, January 19, 2011.

[75] http://www.aipg.org/Seminars/HFMS14/presentations/Dick_Jeffrey.pdf.

[76] https://oilandgas.ohiodnr.gov/portals/oilgas/pdf/YoungstownFAQ.pdf.

[77] Ohio Department of Natural Resources, "Ohio Announces Tougher Permit Conditions for Drilling Activities Near Faults and Areas of Seismic Activity," press release, April 11, 2014, http://ohiodnr.gov/news/post/ohio-announcestougher-permit-conditions-for-drilling-activities-near-faults-and-areas-of-seismic-activity.

[78] Ohio Department of Natural Resources, Division of Oil and Gas Resources Management, Draft Rules and Review (Ch. 1501:9-2-02 OAC), http://oilandgas.ohiodnr.gov/.

[79] Railroad Commission of Texas, Ch. 3. Oil and Gas Division, 39 *TexRegN*8988, November 14, 2014, Texas Register, amending 16 T.A.C. §3.9, §3.46, http://www.sos.state.tx.us/ texreg/pdf/backview/1114/1114adop.pdf.

[80] 39 *TexReg* 8996-9005, 16 T.A.C. §3.9.

[81] 16 T.A.C. §3.9(6)(A)(vi).

In: Deep-Well Injections and Induced Seismicity ISBN: 978-1-63482-557-3
Editor: James B. Fernandez

Chapter 2

MINIMIZING AND MANAGING POTENTIAL IMPACTS OF INDUCED-SEISMICITY FROM CLASS II DISPOSAL WELLS: PRACTICAL APPROACHES*

Underground Injection Control National Technical Workgroup

EXECUTIVE SUMMARY

The U.S. Environmental Protection Agency (EPA) Underground Injection Control (UIC) program regulates injection of fluids related to oil and gas production as Class II injection wells for the protection of underground sources of drinking water (USDWs). Because seismic events from injection have the potential to cause endangerment of underground sources of drinking water, the UIC program director should be aware of that potential and be prepared with response options should seismic events become a concern. Unconventional resources and new technologies, such as horizontal drilling and advanced completion techniques, have expanded the geographic area for oil and gas production activities, resulting in a need for Class II disposal wells in some areas previously considered unproductive.

---

* This is an edited, reformatted and augmented version of a report dated November 12, 2014, issued by the U.S. Environmental Protection Agency, February 2015.

Recently, a number of small to moderate magnitude ($M<5.0$) earthquakes[1] have been recorded in areas with Class II disposal wells related to shale hydrocarbon production. To address the concern that induced seismicity could interfere with containment of injected fluids and endanger drinking water sources, EPA's Drinking Water Protection Division requested that the UIC National Technical Workgroup (NTW) develop a report with practical tools to help UIC regulators address injection-induced seismicity. The Induced Seismicity Working Group (WG) of the NTW developed this report in response, using the existing Class II regulatory framework to provide possible strategies for managing and minimizing the potential for significant injection-induced seismic events.[2] The report focuses on Class II disposal operations and not enhanced recovery wells or hydraulically fractured wells. In formulating the strategies in this report, the NTW conducted a technical literature search and review. Additionally, the NTW evaluated four case examples (in Arkansas, Ohio, Texas, and West Virginia) and considered data availability and variations in geology and reservoir characteristics.

Unconventional production activities and associated larger wastewater volumes have resulted in an increased need for disposal capacity. Some disposal wells handling the increased volumes are located in geographic areas where disposal has not previously occurred. A growing number of disposal wells, some of which are in these new geographic areas, have been suspected of inducing seismicity. Of the approximately 30,000 Class II disposal wells in the United States, very few disposal well sites have produced seismic events with magnitudes greater than M4.0.[3] In addition, EPA is unaware of any USDW contamination resulting from seismic events related to injection-induced seismicity.

Disposal wells are one of a number of historic causes of human activity-induced earthquakes. Others include construction and management of dams and water reservoirs, mining activities, oil and gas production, and geothermal energy production. Evaluation of induced seismicity is not new to the UIC program. The first comprehensive study was completed for the EPA Office of Water over 25 years ago, (Wesson and Nicholson, 1987[4]; finalized as Nicholson and Wesson, 1990). This report is intended to describe for UIC program management the current understanding of induced seismicity within the existing Class II regulatory framework for Class II disposal. The Class II UIC program does not have regulations specific to seismicity but includes discretionary authority that allows additional conditions to be added to the injection permit on a case-by-case basis, along with additional requirements for construction, corrective action, operation, monitoring, or reporting (including closure of the injection well) as necessary to protect USDWs.[5] Legal and policy considerations of Class II regulations, including regulatory revisions, are outside the scope of this technical report. This report is not a guidance document and does not provide specific procedures, but it does provide the UIC Director with considerations for addressing induced seismicity on a site-specific basis, using Director discretionary authority.

The NTW confirmed the following components are necessary for significant injection-induced seismicity: (1) sufficient pressure buildup from disposal activities, (2) Faults of Concern,[6] and (3) a pathway allowing the increased pressure to communicate with the fault. The NTW noted that no single recommendation addresses all of the complexities related to injection-induced seismicity, which is dependent on a combination of site geology, geophysical and reservoir characteristics. An absence of historical seismic events in the vicinity of a disposal well does not provide complete assurance that induced seismicity will not occur; however, this historic absence may be one indicator of induced seismicity if seismic events occur following activation of an injection well. Such a conclusion is based on the assumption that a reliable history of seismic monitoring in the region of the injection well exists. However, the accuracy of such monitoring depends on the robustness of the seismic monitoring network for any given area, along with consideration for how long such a network has been in place. Conclusive proof of induced seismicity is difficult to demonstrate but is not a prerequisite for taking early prudent action to address the possibility of induced seismicity.

The NTW developed a decision model (Figure 1) to inform UIC regulators about site assessment strategies and practical approaches for assessing the three fundamental components. The model begins with considerations for a site assessment dependent on location-specific conditions, because understanding the geologic characteristics of a site is an essential step in evaluating the potential for injection-induced seismicity. Monitoring, operational and management approaches with useful practical tools for managing and minimizing suspected injection-induced seismicity are recommended.

During its review, the NTW also found that the application of basic petroleum engineering practices coupled with geology and geophysical information can provide a better understanding of reservoir and fault characteristics. The multidisciplinary approach offers many ways of analyzing injection-induced seismicity concerns, possibly identifying anomalies that warrant additional site assessment or monitoring. Such an approach would be enhanced by collaborative work between a wide variety of individuals in industry, government, and scientific and engineering research organizations. Consequently, the NTW recommends that future research consider a practical multidisciplinary approach coupled with a holistic assessment addressing disposal well and reservoir behavior, geology, seismology and other appropriate specialty fields of study.

INTRODUCTION

The U.S. Environmental Protection Agency (EPA) Underground Injection Control (UIC) program, authorized by the Safe Drinking Water Act, regulates injection of fluids related to oil and gas production into Class II wells, for the protection of underground sources of drinking water (USDW). There are approximately 30,000 Class II active disposal wells in the United States used to dispose of oil and gas related wastes, many of which have operated for decades. EPA is unaware of any USDW contamination resulting from seismic events related to injection-induced seismicity.[7] Very few of these disposal well sites have produced seismic events with magnitudes[8] greater than M4.0[9]. For example, at the time of this report, there were approximately 2,700 active disposals wells in Louisiana, with no recent significant[10] seismic events occurring as a result of the disposal activities. However, unconventional resources and new technologies, such as horizontal drilling and advanced completion techniques, have increased oil and gas production activities, resulting in a need for new Class II disposal wells in expanded geographic areas.

Disposal wells are just one of a number of historic causes of human activity-induced earthquakes.[11] Other causes may include construction and management of dams and water reservoirs, erection of skyscrapers, mining activities, oil and gas production, geothermal energy production and geologic carbon sequestration.

Enhanced Recovery Injection Wells

Class II injection wells include injection wells used for enhanced recovery as well as those used for oil and gas production wastewater disposal. Injection for enhanced recovery projects generally poses less potential to induce seismicity than wastewater disposal because pressure increases resulting from injection for enhanced recovery are partially offset by nearby production wells. Disposal wells have no offsetting withdrawal and therefore, have a greater potential for pressure buildup. Given the recent seismic activity associated with Class II disposal wells, this report focuses on recommendations to manage or minimize induced seismicity associated with these wells.

Hydraulic Fracturing

Although not the emphasis of this effort, seismicity associated with hydraulic fracturing (HF) was addressed by a review of selected literature sources. HF has a low likelihood of inducing significant seismicity, for reasons explained below.

Unlike wastewater disposal wells where injection occurs for an extended period of time, HF is a short-term event designed to create cracks or permeable avenues in lower permeability hydrocarbon-bearing formations. HF activity is followed by the extraction of reservoir fluids and a decrease in pressure within the formation. Therefore, the "pressure footprint" of a well that has been hydraulically fractured is typically limited to the fracture growth or fracture propagation area (Gidley et al., 1990). In comparison, the "pressure footprint" of an injection well is related to the injection rate, duration of the injection period and transmissibility of the reservoir (Lee et al., 2003). Class II disposal wells typically inject for months or years and generate large "pressure footprints" with no offset production of fluids.

HF is designed to crack the formation to enhance production. Several studies have documented microseismicity ($M<1$) caused by HF (Das and Zoback, 2011; Phillips et al., 2002; Warpinski, 2009; and Warpinski et al., 2012). Studies have also documented numerous examples of small faults encountered during the HF process with microseismicity where magnitudes are below M0 (Maxwell et al., 2011; Warpinski et al., 2008). Recording these very low magnitude seismic events requires the use of downhole seismometers in nearby wells (Warpinski, 2009). Though rare, felt HF-induced seismicity is possible if the HF encounters a Fault of Concern.[12] Documented cases list seismic events up to M3.8 caused by HF communication with Faults of Concern (British Columbia Oil and Gas Commission, 2012; de Pater and Baisch, 2011; Holland, 2011 and 2013; Kanamori and Hauksson, 1992).

Geothermal Injection Wells

A number of informative references exist on induced seismicity and enhanced geothermal systems. These references cover a broad range of seismicity issues and outline many avenues of additional research needed (Hunt and Morelli, 2006; Majer et al., 2007 and 2011). These authors documented the combination of monitoring techniques with adjustment of operational parameters to control seismicity. For example, thermal stress, in addition to pressure buildup, plays a key role in geothermal seismicity and may be applicable to wastewater disposal wells, depending on the temperature of the injected fluids and receiving formation (Perkins and Gonzalez, 1984).

CO_2 Geologic Sequestration

Geologic sequestration of CO_2 requires a Class VI UIC permit. The Class VI permitting process includes assessment of potential induced seismicity. Class VI regulations require a detailed review on a site-specific basis; consequently, Class VI wells were not considered in this report. Some research pertaining to potential seismicity from CO_2 geologic sequestration may be applicable to wastewater disposal.

Directive and Working Group

Revisions to Class II regulations are outside the scope of this technical report. This report is not a policy or guidance document and does not provide an exhaustive list of specific permitting procedures. It provides the UIC Director with considerations for minimizing and managing induced seismicity on a site-specific basis, using Director discretionary authority.

To address the concern that injection-induced seismicity could cause a breach in the containment of injected fluids and endanger drinking water sources, EPA's Office of Ground Water and Drinking Water (OGWDW) Drinking Water Protection Division requested that the UIC National Technical Workgroup (NTW) develop recommendations for consideration by UIC regulators (Appendix A). The UIC NTW consists of UIC staff from each EPA regional office, EPA headquarters, and six state UIC program representatives. The Injection-Induced Seismicity Working Group (WG) of the NTW was formed in June 2011 to spearhead development of a report recommending possible strategies for managing or minimizing significant seismic events associated with induced seismicity in the context of Class II disposal well operations. The WG was comprised of a subset of NTW members and members outside the NTW included for their expertise on the subject matter. A list of the WG members is provided later in this report. Drafts of the report were written by the WG, and finalized based on review by the NTW. Ultimately, the report is a product of the NTW.

REGULATORY AUTHORITIES

This report describes, for UIC regulators, the current understanding of induced seismicity within the existing Class II regulatory framework for Class II disposal. Evaluation of induced seismicity is not new to the UIC program. Some UIC well classes address seismicity with specific regulatory requirements.[13] The Class II UIC program does not have regulations specific to seismicity but rather includes discretionary authority that allows additional conditions to be added to the UIC permit on a case-by-case basis. Examples of this discretionary authority include additional requirements for construction, corrective action, operation, monitoring or reporting; (including well closure) as necessary to protect USDWs.[14] In the included case studies, the UIC Directors used discretionary authority to manage and minimize seismic events.

Potential USDW risks from seismic events could include loss of disposal well mechanical integrity, impact to various types of existing wells, changes in USDW water level or turbidity, USDW contamination from a direct communication with the fault inducing seismicity, or contamination from earthquake-damaged surface sources. However, EPA is unaware of any USDW contamination resulting from seismic events related to injection-induced seismicity.

REPORT PURPOSE

The NTW's task was not to determine if there was a linkage between disposal and seismicity, but if a linkage was suspected, to identify practical approaches the UIC Director may use to minimize and manage injection-induced seismicity. A decision model was developed, which compiles and describes available options and illustrates a process for applying them to manage or minimize possible injection-induced seismicity. The site assessment considerations included in the model were those identified as pertinent by the WG, though other factors may also be appropriate depending on site-specific situations. The decision model also provides operational and monitoring options for managing injection-induced seismicity. It is supported by an extensive literature review and four case histories, which considered earthquake history, proximity of disposal wells to these events, and disposal well behavior.

Many of the recommendations and approaches discussed in this report may be applicable to other well classes. For example, disposal activities also occur in Class I hazardous and nonhazardous wells, various Class V wells, and Class VI wells. The U.S. Department of Energy and the International Energy Agency (IEA) have authored several publications dealing with specific Class V geothermal seismicity issues. The WG reviewed a number of publications as part of the literature survey for this report (Appendix K). Conclusions from some of these reports apply to this Class II injection-induced seismicity project and are referenced within the body of the report.

Injection-Induced Seismicity Project Objectives

The WG analyzed existing technical reports, data and other relevant information on case studies, site characterization and reservoir behavior to answer the following questions:

1) What parameters are most relevant to screening for injection-induced seismicity?
2) Which siting, operating or other technical parameters are collected under current regulations?
3) What measurement tools or databases are available that may be used to screen existing or proposed Class II disposal well sites for possible injection-induced seismic activity?
4) What other information would be useful for enhancing a decision-making model?
5) What screening or monitoring approaches are considered the most practical and feasible for evaluating significant injection-induced seismicity?
6) What lessons have been learned from evaluating case histories?

Working Group Tasks

The UIC NTW was tasked by UIC management with developing a report including technical recommendations to manage or minimize significant levels of injection-induced seismicity.

The UIC NTW utilized the following to address the objectives:

1. Compare parameters identified as most applicable to induced seismicity with the technical parameters collected under current regulations
2. Prepare a decision model
3. Assess applicability of pressure transient testing and/or pressure monitoring techniques
4. Summarize lessons learned from case studies
5. Recommend measurements or monitoring techniques for higher risk areas
6. Analyze applicability of conclusions to other well classes
7. Recommend specific areas for further research needed

Working Group Approach

The WG adopted the following strategy:

1. Summarize geoscience factors and applications
2. Apply petroleum engineering methods
3. Compile and review historical and current scientific literature, including ongoing projects and material associated with upcoming reports on injection-induced seismicity
4. Select and study case examples of Class II brine disposal wells suspected of inducing seismicity and provide a summary of lessons learned for the following areas:
   a. North Texas
   b. Central Arkansas
   c. Braxton County, West Virginia
   d. Youngstown, Ohio

A study of disposal wells in areas with no seismic activity was not performed.

5. Develop a decision model
6. Consult with the U.S. Geological Survey (USGS) seismologists on the potential for using deep stress field measurements and the USGS earthquake information as screening tools (Appendix M)
7. Compare data collected under existing UIC requirements to relevant information needed for assessment of injection-induced seismicity
8. Solicit review by EPA's UIC NTW and subject matter contributors from state agencies, academia, researchers and industry.

Review Process

As noted above, prior to submission to the NTW, the draft report was sent for review to specific subject matter experts and corrections made accordingly. After the NTW passed the report to OGWDW, it was decided to conduct an additional independent peer review.

Peer Review

The OGWDW engaged one of its contractors to facilitate and coordinate an external review of the NTW report. In the process of developing the contract, OGWDW also developed charge questions to guide the reviewers in the areas of desired feedback. With guidance from EPA, the contractor developed a ranked list of about 20 experts. Six reviewers were selected from that list (Table 1), based on their qualifications, including experience with injection-induced seismicity.

Peer Review Charge

The reviewers were asked to focus on four charge questions during their review and to provide expert advice and recommendations on these questions, in addition to providing general comments. The four charge questions, developed by EPA, were as follows:

Basic Mechanism of Injection-Induced Seismicity

The NTW identified three key components contributing to injection-induced seismicity: (1) the presence of a stressed fault, (2) pressure buildup from disposal operations, and (3) a pathway for the increased pressure to communicate from the disposal well to the fault. Do these three key elements capture the causal relationship of disposal-induced seismicity? Please comment on other elements relating induced seismicity to Class II disposal injection that might be useful to consider when developing strategies to minimize or manage injection-induced seismicity.

Table 1. Selected Peer Reviewers

Peer Reviewer	Jeff Bull	Robin McGuire	Craig Nicholson	Kris Nygaard	Heather Savage	Ed Steele
Affiliation	Oil/Gas Industry	Consultant	Academia	Oil/Gas Industry	Academic Laboratory	Oil/Gas Industry and Consultant
Organization	Chesapeake Energy Corporation	Lettis Consultants International, Inc.	University of California, Santa Barbara	Exxon Mobil	Lamont-Doherty Earth Observatory, Columbia University	Swift Worldwide Services
Professional Years	30+	30+	30+	20+	10+	40+

Variety and Validity of Approaches

The NTW identified site assessment considerations along with monitoring and operational approaches for assessing the three key components contributing to injection-induced seismicity. Please comment on the appropriateness of the site assessment considerations identified for assessing the potential for induced seismicity. What other site assessment considerations might be considered? Are the monitoring and operational approaches identified appropriate for minimizing or managing injection-induced seismicity? Are there additional considerations that might be considered to address the key elements contributing to injection-induced seismicity?

Reservoir Engineering Analysis Application

The NTW sought to expand the review of the pressure buildup and pathway components of induced seismicity beyond geosciences. The NTW integrated reservoir engineering analysis into the evaluation of the potential relationship between Class II injection activity and seismicity by using data that is already collected by owners and operators as well as standard evaluation techniques employed in the oil and gas industry. Is the reservoir engineering analysis suggested by the NTW reasonable for identifying anomalies in an effort to minimize or manage injection-induced seismicity? Please identify other analyses (including the type of data needed and benefits and disadvantages of their use) that might be useful for evaluating reservoir behavior during Class II disposal injection.

Recommended Future Efforts for the EPA

Please identify any additional key literature or other data sources that might be useful to ensure a comprehensive understanding of the potential for induced seismicity in the context of Class II disposal wells.

Final Peer Review Follow-up

Once all of the comments were received, the OGWDW requested help in assessing the comments.

The comment review team (team) consisted of EPA Region 6 staff, the past NTW chair and two representatives of OGWDW. The team assessed the comments and divided them into three categories described below; it also developed a strategy to re-engage the NTW for a final review of the report, once updated based on the peer review comments.

The team assessed and tabulated the peer reviewers' comments (Appendix N) according to the relevant section of the report and the commenter. The team then classified each of the comments according to the following categories:

1. Comments requiring no response: These are typically statements or opinions by the commenter.
2. Comments relevant to the topic, but outside the scope of the project: These comments are addressed in more detail in Appendix A. This category of comments was grouped according to the nature of the comment, as described below.
    a. Authority: Comments related to the applicability of EPA authority.
    b. Scope: Comments outside the purpose of the report (providing a practical UIC management tool) including recommendations for policy changes, new regulations, extensive research or additional studies, such as complex proprietary modeling.

c. Clarify: Comments relevant to the topic that were addressed by directing the commenter to the appropriate place in the report or by providing additional UIC program background information.

3. Comments relevant to the topic and within the scope of the project. These comments required revisions to the document.

Additional decisions included the following:

- The original cut-off date for inclusion of case studies (September 2013) was maintained.
- A separate list of the peer reviewers' recommended references was added to Appendix K: Subject Bibliography, excluding non-peer-reviewed articles.
- A new appendix was created to provide responses to all comments grouped in Category 2 (above).
- The following areas were outside the scope of the project and were not incorporated:
  - Adoption of a formal comprehensive risk assessment
  - Specific policy or regulatory requirements
  - Ongoing research, modeling or simulations
  - Basic UIC program discussions

GEOSCIENCE FACTORS RELATED TO INJECTION-INDUCED SEISMICITY

The following paragraphs provide a general overview of the various geoscience aspects relevant to injection-induced seismicity. 0 describes these aspects in greater detail. The three key characteristics related to potential injection-induced seismicity that may lead to fault slippage and associated earthquakes are: (1) an increase in the formation pore pressure from disposal activities; (2) a fault (or zone of multiple faults and fractures) optimally oriented for movement, located in a critically stressed region, of sufficient size, and possessing sufficient accumulated stress/strain, such that fault slip and movement would have the potential to cause a significant earthquake (Fault of Concern); and (3) a permeable avenue (matrix or fracture permeability) allowing the pore pressure increase to reach the fault.

Background

In general, continental oil and gas deposits occur in sedimentary rocks deposited by ancient seas over granitic basement rocks. Basement rocks have been and continue to be subjected to ongoing global tectonic forces. These forces result in fracturing and faulting (fracturing with lateral displacement) and are the origin of the constantly stressed condition of continental basement rocks. Nearly all early cases of suspected injection-induced seismicity felt by humans have involved communication between disposal zones and basement faults. For these reasons, geologic site assessments related to potential injection-induced seismicity should include an analysis of both faults and stress conditions in basement rocks of the disposal well area. Since subsurface geologic stresses are transferred over great distances, fault and stress analyses should encompass a regional area around the disposal well.

Geologic Stress Considerations

Historic seismic activity is an indicator of critical stress in basement rocks. Subsurface stresses are typically not uniform in every direction. The orientation of faults with respect to the principal stresses is a fundamental indicator of which faults are subject to activation from pore pressure increases. Not all faults are Faults of Concern, only those optimally oriented in the subsurface stress field such that an increase in pore pressure can induce movement. Optimal orientation of faults is described in greater detail by Holland (2013). Unfortunately, the principal stress direction may not be readily known to injection well permitting authorities. Some options to help determine the principal stress direction include data on borehole geometry, the World Stress Map (Appendix M: Task 2; Tingay et al., 2006), or consultation with experts, such as state geological surveys or universities. These experts may provide an estimate of the principal stress direction for a particular area as well as information on the location and orientation of known faults in the area.

An additional resource is the Quaternary Fold and Fault Map created by a USGS consortium (Appendix M: Task 1). This map shows all active faults with surface expression that are known to have created earthquakes over M6.0. These faults were defined from the geologic record for the Quaternary age (the last 1.6 million years).

Geophysical Data

Across the United States, the USGS funds or maintains seismic arrays and associated databases that are excellent web-based resources for seismic history assessments. A summary of available databases is provided in Appendix L: . Seismometers in the permanent monitor grid in most of the central and eastern continental United States are spaced up to 200 miles (300 km) apart. With this spacing, the system is capable of measuring events down to approximately M3.0 or M3.5, although in some areas measurement capabilities may extend down to a M2.5. Hypocenter location error for the permanent array averages up to 6 miles (10 km) horizontally and 10,000 to 16,500 feet (3–5 km) vertically. In tectonically active areas such as the continental western margin and New Madrid Seismic Zone, the seismometer spacing is closer, resulting in more accurate earthquake locations (hypocenter by latitude, longitude and depth). Additionally, closer grid spacing generally allows measurement of seismic events of smaller magnitude. Despite the accuracy limitations, USGS or other seismicity databases described in Appendix L and Appendix M are useful tools for initial site assessments. Event information included in databases is periodically updated over time as data are reprocessed. Relocated events are found in later publications and may not be in the seismicity databases.

Communication with Basement Rock

In almost all historic cases, felt injection-induced seismicity was the result of direct injection into basement rocks or injection into overlying formations with permeable avenues of communication with basement rocks. Therefore, the vertical distance between an injection formation and basement rock, as well as the nature of confining strata below the injection zone, are key components of any assessment of injection-induced seismicity. In areas of complex structural history, strata beneath the injection zone may have compromised vertical confining capability due to natural fracturing. Also, faulting in basement rock can extend into overlying sedimentary strata, thus providing direct communication between the disposal zone and the basement rock.

Importance of Porosity and Permeability of Injection Strata

Stratigraphic formations used as disposal zones can have a complex range of porosity types and permeability values. For this report, two fundamental types of porosity are considered; matrix porosity and fracture porosity. Matrix porosity refers to the rock pore spaces, whether formed during deposition or alteration following deposition. Natural fractures in rocks create a second type of porosity referred to as fracture porosity. Fractures can provide preferential flow paths for fluid flow (permeability). Matrix porosity generally is characterized by smaller interconnections and less permeability than fractures, but high matrix porosity offers more storage space, potentially limiting the horizontal extent of pressure distribution. Pressure buildup is more difficult to predict in naturally fractured flow-dominated disposal zones and can extend much farther from the injection well. Most of the case study wells suspected of injection-induced seismicity in this report involved naturally fractured disposal zones.

Petroleum Engineering Applications for Evaluating Induced Seismicity

Petroleum engineering applications have been used for decades in the oil and gas industry to evaluate wells and enhance hydrocarbon production. Petroleum engineering methodologies used in this document adhere to practices and equations commonly presented in petroleum engineering literature. The review of recent injection-induced seismicity literature revealed a lack of a multidisciplinary approach inclusive of petroleum engineering techniques. Additionally, typical Class II disposal permit reviews do not use many of the petroleum engineering analyses available, but such techniques could be useful in evaluating the potential for injection-induced seismicity.

Petroleum engineering methodologies provide practical tools for evaluating the three key components that must all be present for induced seismicity to occur: (1) sufficient pressure buildup from disposal activities, (2) a Fault of Concern, and (3) a pathway allowing the increased pressure to communicate from the disposal well to the fault. Different well and reservoir aspects can be evaluated depending on the methods used. Specifically, petroleum engineering methods typically focus on the potential for reservoir pressure buildup and the reservoir flow pathways present around a well and at a distance, and characterize reservoir behavior during the well's operation. Petroleum engineering approaches enhance geological and seismological interpretations related to the characterization of faults and flow behavior. Some of the case study wells reviewed exhibited specific Hall integral and derivative responses (described further below and in Appendix D) that corresponded to area seismic events. The Hall integral and derivative responses at these wells suggest hydraulic communication with a boundary (i.e., an offset well or fault) at some unknown distance from the well.

The petroleum engineering approach incorporates information typically collected from the permit application (well construction and completion data) and data on injection volumes and pressures reported for compliance purposes during operation of the well. This information is presented in a graphical format to illustrate behavior of the well over time. These graphs are compared to graphs of expected well behavior from various reservoir behavior models to identify anomalous patterns.

Review of operational data can provide a qualitative look at the well behavior. Operational analysis consists of plotting readily available data reported as part of the Class II disposal well permit compliance. These plots include:

- Injection volumes and wellhead pressures
- Bottomhole injection pressure gradient
- Hall integral and derivative

Plotting injection volumes and pressures in an appropriate format along with operating pressure gradients may highlight significant changes in disposal well behavior. The operating gradient plot can indicate whether a disposal well is operating above fracture gradient. The Hall integral and derivative plot utilizes operating data to characterize a well's long term hydraulic behavior by providing a long-term, long distance look into the disposal zone. For example, a decline in wellhead pressures coupled with an increase in volumes injected reflects enhanced injectivity (increased ease of injection), shown by the derivative dropping below the Hall integral, while the derivative trend rising above the integral represents increased injectivity. Changes in Hall integral and derivative trends can represent reservoir heterogeneities (i.e., faults, stratigraphic changes, etc.), changes in completion conditions, reservoir boundaries, and effects of offset wells. Details concerning the application of both the operating gradient and Hall integral and derivative plots are discussed in Appendix D. Both plot types are utilized in the four case studies detailed in Appendices E through H.

Supplemental evaluations may be performed but require data or logs that may or may not be routine for Class II disposal permit activities. These evaluations quantitatively assess potential pathways and potential reservoir pressure buildup and may include the following:

- Step rate tests
- Pressure falloff tests
- Production logs
- Static reservoir pressure measurements

Step rate tests are used to determine the formation parting pressure (fracture extension pressure). The quality of the data analysis is dependent on the amount of pressure data recorded during the test. Pressure falloff tests can provide the completion condition of the well (wellbore skin) and reservoir flow characteristics. Production logs typically include temperature logs, noise logs, radioactive tracer surveys, oxygen activation logs or spinner surveys. These types of logs are used to evaluate the fluid emplacement at the well. Periodic static pressure measurements provide an assessment of reservoir pressure buildup. More details on supplemental testing and engineering evaluations are included in Appendix D.

Review of Scientific Literature

Literature Sources

Injection-induced seismicity has been documented in many reports dating from 1968 through 2013. The WG compiled and reviewed an extensive reference list included in Appendix K. The primary resource was USGS Bulletin 1951 (Nicholson and Wesson, 1990).[15] Induced seismicity is a rapidly expanding area of research. This list is not a complete resource list. Inclusion of an article or website in Appendix K does not reflect NTW's agreement with the conclusion of the article.

Earthquake Reporting

The USGS Advanced National Seismic System (ANSS) comprehensive catalog (ComCat), the largest U.S. database of earthquake events, includes earthquakes from the USGS National Earthquake Information Center (NEIC) and contributing networks. The real-time report and some of the catalogs include the location accuracy of the event. Catalog details may vary, but are an important consideration for induced seismicity analyses. Earthquake catalogs are discussed more fully in Appendix L and Appendix M. USGS, state geologic agencies and universities may also collect and/or host earthquake information on their websites. There may be differences between databases in detection thresholds, as well as inconsistencies in calculated epicenters, depths or magnitudes for each earthquake. Databases may not cover the same geographic regions. It should be noted that the expansion or development of regional seismometer networks may allow measurements of seismic activity at a lower magnitude threshold than previously recorded, creating the appearance of increased seismicity. Event interpretation is discussed more fully in Appendix D.

Possible Causes of Induced Seismicity

Seismicity induced by human activities has been extensively documented. Seismic events have been associated with mining, construction and management of dams and water reservoirs, geologic carbon sequestration, erection of skyscrapers, geothermal energy related injection, oil and gas production activities and disposal wells. Davis and Frohlich (1993), Nicholson and Wesson (1990, 1992), and Suckale (2009, 2010) studied case histories of potential oil- and gas-related induced seismicity across the United States and Canada. Several waste disposal case studies were investigated, including Rocky Mountain Arsenal and Paradox Valley in Colorado, and two locations in far northeastern Ohio (Ashtabula and Cleveland, occurring from 1986 to 2001). Opposing conclusions were drawn on whether the Ohio seismicity was related to injection (Seeber and Armbruster, 1993 and 2004; Gerrish and Nieto, 2003; Nicholson and Wesson, 1990). More recent publications concluded disposal activity induced seismicity in central Arkansas and Youngstown, Ohio (Horton, 2012; Horton and Ausbrooks, 2011; Holtkamp et al., 2013; Kim et al., 2012; Kim, 2013; ODNR, 2012). Disposal activities at the Rocky Mountain Arsenal, Paradox Valley and enhanced recovery at the

Rangely Field, also in Colorado, have been associated with inducing seismicity. Operations at both Colorado facilities began prior to promulgation of federal UIC regulations. Production from the Rangely Field is ongoing.

Several studies concluded that the Rocky Mountain Arsenal seismicity was caused by injection (Davis and Frohlich, 1993; Nicholson and Wesson, 1990 and 1992; Suckale, 2009 and 2010). At the Rocky Mountain Arsenal, the largest three earthquakes, with magnitudes between M5.0 and M5.5, occurred over one year after injection stopped. In March 1962, injection of waste fluids from chemical manufacturing operations at the Rocky Mountain Arsenal was initiated into fractured crystalline basement rock beneath the facility. Initial injection exceeded the formation fracture pressure from March 1962 through September 1963, when the surface pump was removed, leaving injection to continue under hydrostatic pressure. Pumps were once again used for injection from April 1965 through February 1966, when injection ceased. Seismicity started 5 miles (8 km) from the well on April 24, 1962, ranging from M1.5 to M4.4 from 1962 through 1966, with three earthquakes ranging from M5.0 to M5.5 in 1967. Subsequent investigations identified a major fault near the well and showed a direct correlation between increases in bottomhole pressure during injection and the number of earthquakes, using rank difference correlation (Healy et al., 1968; Hsieh and Bredehoeft, 1981; Raleigh, 1972).

From 1969 through 1974, the relationship between seismicity and enhanced recovery injection operations at the Rangely Field in Colorado was studied (Raleigh, 1972; Raleigh et al., 1976). Reservoir pressures were controlled by varying injection rates into enhanced recovery wells and withdrawal rates from production wells within the Rangely Field to determine the relationship between pressure and induced seismicity. Fourteen seismometers deployed throughout the area recorded events ranging from M-0.5 to M3.1, which occurred in clusters in both time and space. Most of these events were below the threshold that is typically felt by humans.[16] Seismometer data and injection pressure and volume data, coupled with modeling, confirmed that earthquakes were induced through an increase in pore pressure. Frictional strength along the fault varied directly with the difference between total normal stress and fluid pressure (Raleigh et al., 1976). Unusual features in this case included measurable response to fluid pressure along one part of the fault, recordable compartmentalization within the reservoir around the fault, and verification that maintaining the reservoir pressure below a calculated threshold stopped the seismicity (Raleigh, 1972; Raleigh et al., 1976). The Rangely Field example illustrates how operational changes can be used to mitigate induced seismicity.

Numerous earthquakes were induced by Class V disposal operations in Paradox Valley, Colorado (Ake et al., 2002 and 2005; Block, 2011; and Mahrer et al., 2005). Seismicity is being managed using intermittent injection periods, injection rate control and extensive seismic monitoring. Additionally, a proposed second Class V disposal well located several miles from the existing well is being evaluated by the U.S. Bureau of Reclamation in response to an expanding area of seismicity. The existing well is required for salinity control for the Delores River and operates above fracture pressure. More information is included in Appendix J.

Disposal wells have been suspected of inducing seismicity in a number of recent cases (USGS, 2013). Verifying the presence of alternative causes of seismicity, such as unusual changes in lake level (Holland et al., 2013; Klose, 2013; El Hariri et al., 2010), is a useful scientific approach.

Determinations of Injection-Induced Seismicity

Nicholson and Wesson (1990) stated that induced seismicity determinations rely on three primary characteristics of earthquake activity:

1. Geographic association between the injection zone and the location of the earthquake
2. Exceedance of theoretical friction threshold for fault slippage
3. Disparity between previous natural seismicity and subsequent earthquakes following disposal with elevated pressures

Davis and Frohlich (1993) developed a practical approach for evaluating whether seismic events were induced by injection based on characteristics similar to those stated by Nicholson and Wesson (1990), e.g., history of previous seismic events, proximity in time and space and comparison of critical fluid pressures. The Davis and Frohlich approach utilizes a series of fundamental questions to evaluate the likelihood of induced seismicity. These questions are outlined below:

1. Are these events the first known earthquakes of this character in the region?
2. Is there a clear correlation between injection and seismicity?
3. Are epicenters near wells (within 3 miles or 5 km)?
4. Do some earthquakes occur at or near injection depths?

5. If not, are there known geologic structures that may channel flow to sites of earthquakes?
6. Are changes in fluid pressure at well bottoms sufficient to encourage seismicity?
7. Are changes in fluid pressure at hypocenter locations sufficient to encourage seismicity?

Although these approaches are qualitative and do not result in positive proof of injection-induced seismicity, they may be useful to UIC regulators as preliminary screening tools. Evaluating causality requires analysis of all important natural and anthropogenic triggers that can disrupt the subsurface stress regimes in proximity to faults in the local area. As such, proof of induced seismicity is difficult to achieve and may be time-consuming but is not a prerequisite for taking early prudent action to address the possibility of induced seismicity.

Note that petroleum engineering techniques used in analysis of oil and gas development were not typically used to evaluate reservoir characteristics potentially associated with induced seismicity in the scientific literature reviewed for this report.

CASE STUDY RESULTS

The WG task was to provide practical tools that the UIC Director could use to assess site conditions prior to developing a plan to minimize and manage seismicity. Case study efforts were directed toward assessments of typical UIC program compliance data and its usability for characterization of injection well behavior and possible correlation with area seismicity. The case studies were not intended to focus on site problems or program administration issues, but rather to determine if practical assessment tools could be developed. The WG also found no indication that the injection wells associated with the case study areas injected outside of the operational boundaries or designated injection zones established by the permit parameters or endangered a USDW.

A total of four geographic areas of suspected injection-induced seismicity were selected by the WG for more detailed evaluation. These case studies were selected from areas where disposal wells were suspected to have caused recent seismic events. Initially, the north Texas, central Arkansas, and Braxton County, West Virginia, areas were selected. The Youngstown, Ohio, area was included later in the project because a disposal well was the suspected cause of a series of seismic events in late 2011. No cases were evaluated where injection-induced seismicity was not suspected.

Initially, the WG identified disposal wells located in the vicinity of recent seismic events in the selected geographic areas. In order to compare well activities to seismic events, a focus area based on a defined radius around the well was used to gather seismic data. Historic seismic events for the cases were derived from six different database catalogs. These external databases are discussed in more detail in Appendix L. A radius of between 5 and 12 miles (8 to 19 km) around each case study well was selected based on the spacing density of the existing seismometers and location of the seismicity in the immediate area of the well. Additionally, there is uncertainty regarding the depth to the hypocenter.

The specific strategies used by the WG for evaluating the cases included engaging researchers who had studied two of the cases, reviewing available geologic structure maps, acquiring specific injection well data from the four state regulatory agencies and communicating with a well operator. A petroleum engineering analysis, based on the collected well data, was also performed on each case study well. Additional geoscience background and the results of EPA's petroleum engineering analysis on these cases are discussed in greater detail in the appendix specific to each case study (Appendix E, Appendix F, Appendix G, and Appendix H).

Each case is discussed below through a background summary of the seismic activity and a description of how the case was evaluated by the WG. A summary of the common characteristics and lessons learned from the case studies is included following the case study summaries.

North Texas Area

Several small earthquakes occurred in the central part of the Dallas-Fort Worth metroplex near the Dallas-Fort Worth (DFW) International Airport on October 31, 2008, and near the town of Cleburne on June 2, 2009. Both areas are located in north central Texas, in the eastern portion of the Barnett shale play. Prior to 2008, no earthquakes had been reported within 40 miles (64 km) of the DFW and Cleburne case study areas. Although Barnett shale hydrocarbons were discovered in Wise County in 1981, extensive drilling into the Barnett shale began only in the late 1990s with the advancement of horizontal drilling and well completion technologies. Disposal wells are the primary management approach for handling the wastewater associated with increased drilling activities. As of January 23, 2012, there were 195 UIC permits for commercial disposal wells in the 24-county area, only 2 of which were permitted in 2012, and not all of which were active.[17]

The Railroad Commission of Texas (RRC) standard UIC permit application package incorporates some site data and well construction and completion information along with other supporting documentation to demonstrate the protection of USDWs (Johnson, 2011). Site documentation reviewed by the WG included surface maps, location plats, disposal depths and inventory of offset wells within the area of review. Well construction details provided to the state include well specifics (e.g., casing, cement information, perforations, and completion information) and disposal conditions (e.g., disposal zone, maximum allowable injection rate and surface pressure). In addition, an annual report filed by the operator provides monthly injection volumes and pressure data. WG review of the annual injection reports indicated that the study area wells operated within the permitted pressure limits. One of the Cleburne area disposal wells was dually permitted as a Class II and Class I disposal well by different regulatory agencies. UIC Class I well requirements include conducting annual falloff tests. These tests provided reservoir characteristics and pressures for compliance with the Class I well permit and were not required in response to area seismicity. The WG reviewed the available falloff tests that confirmed the Ellenburger disposal interval was naturally fractured.

Following the 2008 and 2009 events, the RRC identified active disposal wells in the area for further evaluation due to the wells' proximity to the epicenters of seismic events and the absence of seismicity prior to initiation of disposal. RRC opened a dialogue with the operators of the suspect disposal wells, resulting in the voluntary cessation of injection for two wells, one in the DFW area and one in the Cleburne area, in August 2009 and July 2009, respectively. Since the two wells were shut-in, the frequency of seismic events in the immediate focus area, as reported by the USGS website, has substantially decreased.

The RRC subsequently reviewed its permit actions for these wells and other wells in the area in an effort to determine if the activity could have been predicted. No indications of possible induced seismicity were found in these reviews. RRC also inspected the area to verify there were no resulting public safety issues from these events. In follow-up, the RRC consulted with industry representatives, along with researchers at the Texas Bureau of Economic Geology, Southern Methodist University and Texas A&M University. The

RRC continues to monitor developments and research related to injection-induced seismicity.

However, later seismic activity in the DFW area was reported in Janská and Eisner (2012) and new episodic seismic events have occurred in other areas around Cleburne since the initial case study. Reviewing the multidisciplinary findings, available WG flow analysis supports cyclic radial flow followed by linear, fracture flow in the Ellenburger, a karstic carbonate disposal zone. There is a possibility that a few of the wells may have unintentionally created additional fracturing at the operating disposal pressures. Additionally, there appears to be a pattern of repeating cycles of decreased ability to inject followed by enhanced ease of injection, with the decreased injectivity corresponding to seismic events.

More details on this case study are available in Appendix E.

Central Arkansas Area

From 2009 through 2011, a series of minor earthquakes occurred in the Fayetteville shale play near the towns of Guy and Greenbrier in Faulkner County, Arkansas. Regionally, the Enola area, located approximately 9 miles (14.5 km) southeast of Greenbrier, experienced a swarm of earthquakes starting in 1982 (Ausbrooks and Doerr, 2007).

The Arkansas Oil and Gas Commission (Commission) has members appointed by the governor of Arkansas. The Commissioners oversee the state oil and gas agency, also called the Arkansas Oil and Gas Commission (AOGC).

The AOGC standard UIC permit application package incorporates site assessment, well construction and completion information, along with other supporting documentation to demonstrate the protection of USDWs. Site assessment documentation includes surface maps, location plats, disposal depths, and inventory of offset wells within the area of review. Some permit applications contain detailed geologic information, such as a narrative, structure map, type log and additional interpretive data. Well construction details provided to the state include well specifics (e.g., casing, cement information, perforations, and completion information), and disposal conditions (e.g., disposal zone and maximum allowable injection rate and surface pressure). In addition, an annual report filed by the operator provides monthly injection volumes and pressure data. For one disposal well closest to the Enola area earthquakes, the AOGC also requires pressure falloff testing, additional seismic monitoring and intermittent injection during the permitting process. WG review of the annual injection reports indicated that the Enola area well operated within the permitted pressure limits.

In October 2009, 3.5 months after injection commenced, earthquake activity began in the immediate Greenbrier area. To investigate the earthquakes, the AOGC worked with the Arkansas Geological Survey (AGS) and the University of Memphis Center for Earthquake Research and Information (CERI), and additional seismographs were deployed. In December 2010, following increased frequency and higher magnitude earthquakes, the Commission established a moratorium on the drilling of any new Class II disposal wells in an area surrounding the immediate vicinity of the increased seismic activity. The Commission also required the operators of the seven existing Class II disposal wells operating in the moratorium area to provide hourly injection rates and pressures on a bi-weekly basis for a period of 6 months, through July 2011. During the moratorium period, the AGS and CERI analyzed the injection data and seismic activity to determine if there was a relationship.

In late February 2011, following a series of larger magnitude earthquakes, the operators of three disposal wells nearest to the seismic activity voluntarily terminated well operations prior to the issuance of the Commission cessation order issued on March 4, 2011. In July 2011, following the conclusion of the moratorium study, the Commission established a revised permanent moratorium area in which no additional Class II disposal wells would be drilled and required four of the original seven disposal wells to be plugged. The revised moratorium area was based on the trend of the Guy-Greenbrier fault, which the Commission determined as the probable cause of the seismic activity. The operators of three of the wells voluntarily agreed to plug the subject disposal wells and were consequently not parties to the July 2011 hearing heard by the Commissioners. Following the July 2011 Commission hearing, the AOGC issued an order to the operator of the fourth disposal well to plug that well. The order of the Commission issued in July 2011 became a final administrative regulation on February 17, 2012.

Since July of 2011, the AOGC, AGS and CERI continue to monitor disposal well operations and seismic activity. Additional seismic monitoring equipment has been purchased to facilitate the creation of an "early warning" system for emerging seismic activity, thereby allowing more time to develop appropriate responses.

Reviewing the multidisciplinary findings, operational data analysis indicated cycles of upward and downward shifts in both the Hall integral and derivative trends on the various plots for the four disposal wells with adequate monitoring history. As in other case studies, the upward shifts had at least some correspondence to area seismic events. The cyclic tandem plot patterns, when considered in conjunction with the area geology, embedded pressure transient tests, and the operating gradient plots, likely reflect a combination of reservoir rock heterogeneities, fracturing occurrence in the wells in the form of enhanced injectivity, and interaction with reservoir boundaries such as a fault.

More details on this case study are available in Appendix F.

Braxton County, West Virginia

In April 2010, a series of earthquakes ranging in magnitude from M2.2 to M3.4 began in Braxton County, West Virginia. This area had previously experienced a M2.5 earthquake in 2000. Braxton County is located on the eastern edge of the Marcellus shale play, and drilling in this area began in 2006. In March 2009, a nearby Class II disposal well began injecting Marcellus oil and gas production wastewater into the Marcellus formation.

The West Virginia Department of Environmental Protection (WVDEP) Office of Oil and Gas standard UIC permit application package incorporates site assessment, well construction and completion information, along with other supporting documentation to demonstrate the protection of USDWs. The permit application for the well of concern contained detailed geologic information, such as an isopach and structure map. Site assessment documentation included surface maps, location plats, disposal depths, and inventory of offset wells within the area of review. Well construction details provided to the state included well specifics (casing, cement information, perforations and completion information) and disposal conditions (interval, rate and maximum pressure requested). The results of a step rate test were also included with the permit information. In addition, an annual report filed by the operator provides monthly injection volumes and pressure data. WG review of the annual injection reports indicated that the well operated within the permitted pressure limits. The data reported by the operator indicated that the well did not operate continuously.

In response to the seismic activity, the WVDEP reduced the maximum injection volume in September 2010. No additional earthquakes were recorded in the area after this restriction was enacted, until January 2012, when a M2.8 earthquake occurred. In response, the WVDEP further reduced the allowable monthly disposal volume by half the permitted value and researched the geologic structure of the area. The WVDEP and the WG found no conclusive evidence linking the cause of the seismicity to the disposal well.

In February 2012, WVDEP began requiring UIC permit applications to include detailed geologic information specifically to identify subsurface faults, fractures or potential seismically active features. This additional information requirement includes public or privately available geologic information, such as seismic survey lines, well records, published academic reports, government reports or publications, earthquake history, geologic maps or other like information to determine the potential for injection to lead to activation of fault features and increase the likelihood of earthquakes.

Reviewing the multidisciplinary findings, operational analysis of the single disposal well injecting into the Marcellus shale indicates a hydraulic response. Based on the tandem plot analysis, a reservoir boundary (or boundaries) such as a fault, a pinch out, or possibly the limits of fracture stimulation (effectively the limits of permeable rock) was encountered.

More details on this case study are available in Appendix G.

Youngstown, Ohio

Starting on March 17, 2011, a series of 12 small magnitude seismic events occurred in Mahoning County in and around Youngstown, Ohio, culminating in a M4.0 event on December 31, 2011. Evidence suggested that the newly permitted, Northstar 1 Class II saltwater disposal well was the cause of the seismic activity, and the injection well was voluntarily shut down a day before the M4.0 event. The Northstar 1 injection well had been permitted as a deep stratigraphic test well and was drilled to a depth of 9,184 feet into the Precambrian basement rocks in April of 2010. On July 12, 2010, the Northstar 1 was issued a Class II saltwater disposal permit, and injection operations commenced on December 22, 2010.

The first Class II saltwater disposal well was permitted in Mahoning County in 1985, and eight more wells were converted to Class II injection between 1985 and 2004. These Class II injection wells utilized depleted oil and gas zones or were plugged back to shallower, non-oil or gas geologic formations for disposal. Injection was predominantly for disposal of production brine associated with conventional oil and gas operations. With the development of the unconventional shale plays in Pennsylvania and the lack of disposal in Pennsylvania, there was a need for additional disposal operations.

To accommodate some of this need, five commercial disposal wells (Northstar 1 through 4, and 6) were permitted and drilled in Mahoning County, Ohio.

Historically, seismic monitoring in Ohio has been sporadic, and seismic events have not been accurately determined. In 1999, the Ohio Seismic Network (OSN) was established with 6 stations, and there were 24 seismic stations in operation in 2011. The seismometer at Youngstown State University was added to the OSN in 2003. Due to the continued seismic events occurring around the Youngstown area and near the Northstar 1 injection well, four portable seismic units were deployed on December 1, 2011, by Lamont-Doherty Earth Observatory. This portable array allowed more accurate identification of seismic events. After the M4.0 event on December 31, 2011, the governor of Ohio placed a moratorium on other deep injection wells within a 7 mile radius of the Northstar 1 and put a hold on the issuance of any new Class II saltwater injection well permits until new regulations could be developed.

There is a seismically active zone in western Ohio and several episodically active faults 20 and 40 miles away from Youngstown (Baranoski, 2002 and 2013). Prior to the earthquakes recorded in 2011, the only known deep-seated fault was mapped approximately 20 miles (32 km) away from the seismic activity, based on a Pennsylvania Geological Survey report (Alexander et al., 2005). The vast majority of all historic and current seismic activity in Ohio occurs within the Precambrian basement rocks.

Due to the lack of deep geological information available for the Mahoning County area, a deep Precambrian basement fault in close proximity to the Northstar 1 went undetected. This fault was confirmed through evaluation of geophysical logs from the offset deep disposal wells and an interpreted seismic line.

According to the *Preliminary Report on the Northstar 1 Class II Injection Well and the Seismic Events in the Youngstown, Ohio, Area* (Ohio Department of Natural Resources, 2012), data suggest seismicity in the Mahoning County area is related to Class II disposal. The Northstar 1 was drilled 200 feet into the Precambrian basement rock. The Ohio Department of Natural Resources (ODNR) report also suggests that pressure from disposal activities may have communicated with the Fault of Concern located in the Precambrian basement rock. The ODNR now prohibits the drilling of Class II injection wells into the Precambrian basement rock and has enhanced the standard UIC permit requirements[18] to facilitate better site assessment and collection of more comprehensive well information. The additional permit requirements include the following options 'as deemed necessary' and are reviewed on a well-by-well basis: pressure falloff testing, geologic investigation to identify faulting in the immediate vicinity of the well, a seismic monitoring plan or seismic survey, comprehensive suite of well logs, an initial bottomhole pressure measurement and a radioactive tracer or spinner survey. Additional operational controls[19] consist of: daily injection volume and pressure monitoring; an automatic shut-off system; and monthly monitoring of annular pressure.

In late 2012, ODNR also implemented a proactive approach to seismic monitoring around deep Class II disposal wells in Ohio and purchased nine portable seismic units to bolster earthquake monitoring capabilities. All nine portable seismic units are in operation, and ODNR has been monitoring these seismic stations in real-time since late 2012. Additionally, two disposal well operators have installed their own portable seismic arrays around two new wells that ODNR is also monitoring in real-time.

Reviewing the multidisciplinary findings led to the following summary: the Northstar 1 injection well was completed into an approximately 900 foot openhole interval that crossed multiple formations, including faulted basement rock. A production log indicated flow likely occurred into an openhole interval above the basement rock; however, the entire completion interval was exposed to the well's operating pressure. The tandem plot indicated, as in the other case studies, several cycles of decreasing and increasing ease of injectivity, with some correspondence between seismic events, and a portion of the cycles displaying decreasing injectivity (Hall derivative upswings).

More details on this case study are available in Appendix H.

Common Characteristics, Observations, and Lessons Learned from Case Studies

The case studies highlighted in this report provided important lessons and observations as well as common characteristics for wells suspected of inducing seismicity. The lessons learned provided a basis for the decision model as well as the approaches for minimizing and managing induced seismicity. The case study common characteristics and observations contributed to the site conditions component of the decision model. Common characteristics, coupled with key case study observations and the lessons learned, are summarized below.

Common Characteristics and Observations

The common characteristics and observations represent those aspects noted by the WG across multiple case studies.

- Petroleum engineering analysis indicated some correspondence between disposal well behavior and seismicity (all case study areas).
- The magnitude of the earthquakes may increase over time as observed in some case studies (central Arkansas, Ohio and West Virginia).
- Injection into fractured disposal zones directly overlying or connected to basement rock may be more vulnerable to injection-induced seismicity (Arkansas and Ohio case study areas).
- Deep disposal wells were in direct communication or suspected to be in hydraulic communication with basement rocks and Faults of Concern, as in the central Arkansas and Ohio examples. Disposal commonly occurred into disposal zones with naturally fractured reservoir characteristics, as in the central Arkansas and north Texas case study examples.
- Operational analysis of disposal rates and pressures on case study wells showed multiple incidences of repeating cycles of decreased ability to inject followed by enhanced ease of injection, with the decreased injectivity corresponding to seismic events (all case study areas).
- Operating wells below fracture pressure avoids or minimizes fracture propagation. Determination of appropriate operating conditions may require actual testing, such as a step rate test, to measure the formation parting pressure, or conducting an operational analysis for indication of enhanced injectivity.

Lessons Learned

The following key lessons were learned from the case study reviews:

- Initiating dialogue with operators can enhance cooperation, resulting in early voluntary action from operators, including well shut-in, or acquisition of additional site data.
    - Initiating dialogue between the operator and UIC regulator resulted in the voluntary shut-in of some suspect disposal wells (north Texas, central Arkansas and Ohio).
    - In two instances, an operator showed a proprietary 3-D seismic interpretation to the permitting authority, revealing a deep-

seated fault (north Texas and central Arkansas).

- Analysis of existing operational data may provide insight into the reservoir behavior of the disposal zone (all case study areas).
    - Hall integral and derivative plots may indicate a no-flow boundary, such as a fault plane or stratigraphic pinch out, at a great distance.
    - Hall integral and derivative plots may illustrate enhanced ease or increased difficulty of injection.
- Enhanced injectivity could represent injection-induced fracturing, opening or extension of natural fractures, higher pressures allowing fluid flow into lower permeability portions of the formation, or encountering an increased permeability zone at distance (all case study areas).
- Acquisition of additional data may provide an improved analysis.
    - Additional site characterization may be beneficial:
        - Demonstrating a confining layer between the disposal zone and basement rock, and structural interpretation does not indicate faults extending into basement rock.
    - Increased recording of operational parameters can improve the quality of the operational data analysis.
        - Increased frequency of monitoring for permit parameters improved the operational analysis (central Arkansas and Ohio).
- Conducting a falloff test can further refine the reservoir characterization.
    - Fractured flow behavior was confirmed from the falloff test analyses for the Ellenburger disposal zone in a Cleburne area well (north Texas).
- Engaging external geophysical expertise may allow determination of a more accurate location (x,y,z) of the active fault and stress regime, through reinterpretation or increased seismic monitoring.
    - Especially useful when earthquake event magnitudes increased over time (central Arkansas, Ohio and West Virginia).
- Lack of historic seismic events may be a function of lack of seismic activity, seismic activity below recordable levels, or epicenters away from population centers.
- Existing seismic monitoring stations are generally insufficient to

pinpoint active fault locations; more sensitive and better located monitoring systems are needed to accurately identify active faults and detect smaller events.

- o Installation of additional stations resulted in reliable identification of active fault locations (central Arkansas and DFW airport area of north Texas).
- o Epicenters of recorded events are scattered, due to an insufficient number of network stations in proximity to the activity (West Virginia).

- Seismic event data is periodically updated.
  - o During preparation of this report the seismicity data were downloaded on different dates, with many of the initial events later revised or deleted.
    - Deletions typically occur between the first event report and entry into the catalog (NEIC or ComCat).
    - Revisions cover 3-D location as well as magnitude.
      - Several of the catalogs have added a revision date to their entries to help identify such changes.
- Seismic event data may be reprocessed, resulting in relocation of the event.
  - o Fine-tuned relocation is possible when a sufficiently detailed velocity model is developed.
  - o Relocated events are found in later publications and may not be in the catalogs.
- A multidisciplinary approach helped to minimize and manage induced seismicity at a given location (all case study areas).
  - o State geological survey or university researchers provided expert consultation, facilitated installation of additional seismometers and provided a clearer understanding of the deep-seated active faulting (north Texas and central Arkansas).
- Director discretionary authority was used to solve individual site-specific concerns:
  - o Directors used authority to acquire additional site information, request action from operators and prohibit disposal operations. For example, directors used the following approaches:
    - Increased monitoring and reporting requirements for disposal well operators to provide additional

operational data for reservoir analysis (central Arkansas).

- Required one operator to install a seismic monitoring array prior to disposal as an initial permit condition (central Arkansas).
- Plugged or temporarily shut-in suspect disposal wells linked to injection-induced seismicity while investigating or interpreting additional data (all case study areas).
- Defined a moratorium area prohibiting Class II disposal wells in a defined high-risk area of seismic activity (central Arkansas).
- Decreased allowable injection rates and total monthly volumes in response to seismic activity (West Virginia).

DECISION MODEL

The primary objective of the WG was to develop a practical tool, the decision model, for the UIC Director to consider in minimizing and managing injection-induced seismicity potentially associated with new or existing Class II disposal wells. The decision model is specifically designed for Class II disposal wells. However, the UIC Director should also consider other causative factors, such as lake level changes or different types of area operations (mining, production activities, etc.). As mentioned previously, the three key components behind injection-induced seismicity are (1) sufficient pressure buildup from disposal activities, (2) a Fault of Concern, and (3) a pathway allowing the increased pressure to communicate from the disposal well to the fault. All three components must be present to induce seismicity. The decision model was designed to identify the presence of any of the three key components. Based on the historical successful implementation of the UIC program, the decision model would not be applicable to the vast majority of existing Class II disposal wells since most are not associated with seismic activity. Use of the decision model is predicated on UIC Director discretionary authority. Federal UIC regulations do not specifically address risk consequences associated with seismicity, but allow the UIC Director discretion to ensure protection of USDWs.

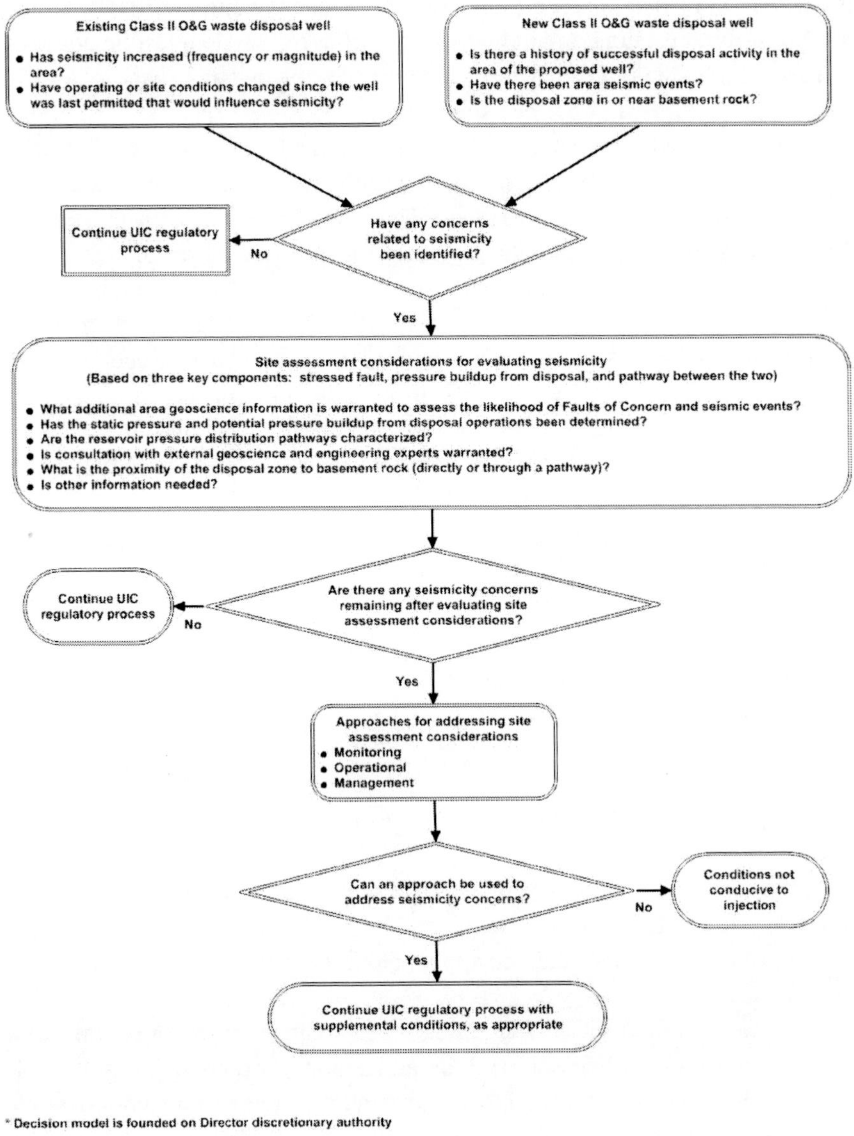

Figure 1. Injection-Induced Seismicity Decision Model for UIC Directors* (Based on the decision model discussion in Appendix B).

The decision model incorporates a site assessment consideration process addressing reservoir and geologic characteristics related to the three key components. The decision model provides the UIC Director with specific site assessment considerations and approaches to identify and address seismicity criteria for both existing and new disposal wells. No single question addresses all the considerations needed to evaluate a new or existing disposal well. If issues are identified, the decision model provides specific operational, monitoring and management approaches as options for addressing the issues.

The diagram of the decision model, Figure 1, is followed by a discussion of considerations for site assessment. The "area" referenced in the decision model is a geographic area with the extent determined by the Director using expertise about the site circumstances. Issues identified through the site assessment consideration thought process are then addressed, as needed, by a combination of operational, monitoring and management approaches. These options were identified by the WG from petroleum engineering methods, literature reviews, analyses of the case studies, and consultations with researchers, operators and state regulators. A more detailed discussion of the decision model is included in Appendix B.

The decision model (Figure 1) contains three symbols that represent the following:

- Bubble – thought process
- Diamond – decision point
- Rectangle – outcome

Existing or New Class II Disposal Well

The decision model was designed to address seismicity concerns related to new or existing disposal wells. Below are the different scenarios. Different site assessment considerations may be applicable to each scenario.

1. An existing disposal well operating in a zone with historical injection
2. An existing disposal well in an area not experiencing seismicity, where the operator requests a substantial increase to injection volumes or pressure
3. A new disposal well in a disposal zone or area where little or no disposal activity has previously occurred, with or without seismic activity

Scenario (1) may not warrant further site assessment based on successful historical operations, while scenarios (2) or (3) may warrant additional site characterization consideration, especially if the well is located in a region with possible Faults of Concern.

Have Any Concerns Related to Seismicity Been Identified?

An UIC Director who does not identify any injection-induced seismicity concerns may exit the decision model and continue through the normal UIC regulatory process; otherwise, a continuation through the model for further site assessment considerations may be warranted.

Site Assessment Considerations

Site assessment considerations identify and help the UIC Director evaluate any specific site characteristics that raise potential issues regarding injection-induced seismicity. Uncertainties about any one of the three key components may warrant collection or review of additional data within the site assessment consideration process.

Site assessment considerations may pertain to information from permit applications or post-approval permit monitoring data. Site assessment considerations may include aspects of both geosciences and petroleum engineering, so a multidisciplinary approach is advantageous. Details about the decision model diagram and its associated site assessment considerations are provided in Appendix B. Site assessment considerations determined to be relevant for the decision model were the following:

- What additional area geoscience information is warranted to assess the likelihood of Faults of Concern and seismic events?
- Has the static pressure and potential pressure buildup from disposal operations been determined?
- Are the reservoir pressure distribution pathways characterized?
- Is consultation with external geoscience and engineering experts warranted?
- What is the proximity of the disposal zone to basement rock (directly or through a pathway)?
- Is other information needed?

Are There Any Seismicity Concerns Remaining after Site Assessment?

An UIC Director who does not identify any injection-induced seismicity concerns following a more detailed site assessment may exit the decision model and continue through the normal UIC regulatory process. When an injection-induced seismicity concern is identified, the Director may determine an approach to address the concern. The site assessment considerations are intended to guide the Director in selecting operational, monitoring and management approaches that are appropriate for addressing induced seismicity issues.

Approaches for Addressing Site Assessment Issues

There are a number of approaches available to manage and minimize significant seismic events. These can be broadly categorized as operational, monitoring and management approaches. An operational approach may include, for example, restricting the maximum allowable injection rate or pressure. A monitoring approach may necessitate collection of additional monitoring data, for example, operational pressures, additional seismic monitoring or pressure transient well testing. A management approach supports a proactive approach for prompt action following seismic events and promotes agency, operator and public interaction. The UIC Director determines which, if any, approaches are important, depending on site-specific considerations. Details about the approaches for addressing issues associated with the site assessment considerations are provided in Appendix B.

Can an Approach Be Used to Successfully Address Seismicity Concerns?

Where the UIC Director does not identify a suitable approach for addressing seismicity concerns, conditions may not be suitable for disposal operations at that location. If monitoring, operational or management approaches provide the required level of protection, the Director may condition the permit accordingly or use discretionary authority to require the desired approaches without revoking the permit.

RESEARCH NEEDS

The WG did not exhaust all avenues with respect to research on the value of petroleum engineering approaches. An abundance of research describing seismology and geomechanical behavior in the form of physical rock properties exists, although studies that combined petroleum engineering and geoscience approaches could not be found by the WG. The WG recommends future practical research using a multidisciplinary approach and a holistic assessment addressing disposal well and reservoir behavior, geology and area seismicity. Such an approach would benefit from combined expertise in geology, petroleum engineering, geophysics and seismology, which may not be available through one entity. For example, areas of expertise should include, but may not be limited to structural and stratigraphic geology; rock mechanics (aka geomechanics); seismology; reservoir characterization; reservoir fluid flow mechanisms; and disposal well construction, completion and performance.

The WG employed Hall plots for the petroleum engineering analysis because regulators may perform these analyses using widely available spreadsheet software and routinely obtained program data. However, other petroleum engineering evaluations are also available that may be applicable, if converted to incorporate injection conditions. The WG identified correspondence between injection well operational characteristics and seismic events in some of the case study wells using Hall plots. Future research is needed to explore other simple engineering techniques that could be used to analyze potential correlations between disposal well operational long-term hydraulic behavior and earthquake events. One of the key outcomes of such a research project would be a practical set of methodologies to assess operating data using injection well permit reporting data normally acquired for existing UIC permits.

To clarify the meaning of the injectivity patterns observed in the case study wells, a comparison of typical injectivity responses for disposal wells in different fractured and unfractured formations would be invaluable. There are a host of variations on this theme, where additional information is needed in order to identify whether a response is associated with a single cause or stems from multiple sources. This information includes such things as formation character, offset disposal well interaction, proximity to a fault, and fracture initiation. A correlative study analyzing whether or not microseismicity accompanies the disposal would help to clarify the risk aspect. Where seismic responses appear, understanding the timing of disposal operations and the apparent response would be an important addition to the UIC knowledge base.

There is also a need for research related to geologic siting criteria for disposal zones in areas with limited or no existing data. The geologic and geophysical study could focus on new stratigraphic horizons that could serve as disposal zones in these areas, the nature of subsurface stresses in basement rocks of these areas, and a more detailed regional geological assessment of basement faults. If sufficient earthquake catalog data are available, additional research to devise a statistical analysis to relate Class II disposal wells operating parameters to induced seismicity would be useful.

Recommendations for Minimizing or Managing Injection-Induced Seismicity

The WG found that no single recommendation addresses all the complexities related to managing or minimizing injection-induced seismicity. Recommendations included in this report were derived from a combination of WG expertise, case studies, consultations with outside experts and data from literature reviews. Recommendations from the outcome of the decision model can be divided into three technical categories (site assessment considerations, operational and monitoring) and a management component. An early step in the induced seismicity evaluation process is to conduct a preliminary assessment. Based on the preliminary assessment and additional site assessment considerations, further operational, monitoring and management approaches may be warranted. The complete discussion of the decision model is located in Appendix B.

Preliminary Assessment of Existing or New Oil and Gas Waste Disposal Wells

- Assess disposal history for correlation with area seismicity.
- Review area seismicity for increases in frequency or magnitude.
- Identify changes in disposal well operating conditions that may influence seismicity.
- Determine the depth to basement rock and potential connectivity to the disposal zone.

Site Assessment Considerations

Site assessment considerations were developed to identify and evaluate specific site characteristics that may represent potential issues for injection-induced seismicity. Many geologic and petroleum engineering considerations for site characterization are not part of the typical permit application process. Additional data collection or review of additional data may be warranted. Possible site assessment activities are shown below:

- Evaluate regional and local area geoscience information to assess the likelihood of activating faults and causing seismic events.
- Assess the initial static pressure and potential pressure buildup in the reservoir.
- Review the available data to characterize reservoir pathways that could allow pressure communication from disposal activities to a Fault of Concern.
- Consult with external geoscience or engineering experts as needed to acquire or evaluate additional site information.
- Determine the proximity of the disposal zone to basement rock.
- Consider collecting additional site assessment information in areas with no previous disposal activity and limited geoscience data or reservoir characterization, prior to authorizing disposal.

Approaches

Possible operational, monitoring and management approaches follow to address seismicity concerns that may arise from the site assessment evaluation. Several proactive practices were identified for managing or minimizing injection-induced seismicity. The applicability and use of any of these approaches should be determined by the Director.

Operational Approach

- Conduct a petroleum engineering analysis of operational data on wells in areas where seismicity has occurred, to identify potential correlation.
- Conduct pressure transient testing in disposal wells suspected of causing seismic events to obtain information about injection

zone characteristics near the well.

- Perform periodic static bottomhole pressure monitoring to assess current reservoir pressures.
- Modify injection well permit operational parameters as needed to minimize or manage seismicity issues. This may require trial and error. Examples of modifications may include the following:
    - Reduce injection rates, starting at lower rates and increasing gradually.
    - Inject intermittently to allow time for pressure dissipation, with the amount of shut-in time needed being site-specific.
    - Separate multiple injection wells by a larger distance for pressure distribution since pressure buildup effects in the subsurface are additive.
    - Implement contingency measures in the event seismicity occurs over a specified level.
- Operate wells below fracture pressure to maintain the integrity of the disposal zone and confining layers.
- Perform annular pressure tests and production logging if mechanical integrity is a concern.

Monitoring Approach

- Increase frequency of monitoring for injection parameters, such as formation pressure and rates, to increase the accuracy of analysis.
- Monitor static reservoir pressure to evaluate pressure buildup in the formation over time.
- Install seismic monitoring instruments in areas of concern to allow more accurate location determination and increased sensitivity for seismic event magnitude.
- Increase monitoring of fluid specific gravities in commercial disposal wells with disposal fluids of variable density since the density impacts the bottomhole pressure in the well.

Management Approach

- For wells suspected of induced seismicity, take early actions, such as acquiring more frequent reports of injection volumes and pressures, reducing injection rates, and/or increasing seismic monitoring, rather than waiting on definitive proof of the causal relationship. Engage the operators early in the process, especially in areas that are determined to be vulnerable to injection-induced seismicity.

- Engage external multidisciplinary experts from other agencies or institutions. For example, Directors may utilize geophysicists to interpret or refine data from seismic events for accuracy and stress direction.
- Provide training for UIC Directors on new reservoir operational analysis techniques to help them understand the spreadsheet parameters.
- Employ a multidisciplinary team for future research to address possible links between disposal well and reservoir behavior, geology and area seismicity.
- Include a seismic threshold as a condition of the permit describing action to be taken in the event of initiation of or increase in seismic events. Thresholds could be based on the magnitude or frequency of events.
- Develop public outreach programs to explain the complexities of injection-induced seismicity.

REPORT/END PRODUCT TASK RESULTS

EPA requested that the NTW output include a specific list of elements in the final report (Appendix A). This list is repeated below, with the corresponding section of this report summarizing the results listed immediately below the item. (Report locations are italicized.)

1. Comparison of parameters identified as most applicable to induced seismicity with the technical parameters collected under current regulations

A point-by-point comparison is not possible as program requirements are widely variable across the various EPA regions and state agencies. The most commonly requested disposal permit parameters found to be useful in addressing potentially induced seismicity include accurate reporting[20] for the following:

a. All available disposal formation data with respect to flow characteristics and continuity; i.e., static pressure, permeability, normal flow pattern (homogenous or linear) and potential disruptions to flow path (stratigraphic or structural)

b. Annual reports of injection volumes and pressures (average and maximum); monthly is more useful than quarterly; daily is needed for more refined analysis

2. Decision-making model—conceptual flow chart

Figure 1 under *Decision Model: Site Assessment Considerations* and at the end of *Appendix B*

a. Provide strategies for preventing or addressing significant induced seismicity. (Note that prevention of earthquakes may not be possible where faults are critically stressed.)

i. Recommendations for Minimizing or Managing Injection-Induced Seismicity; and first subheading

ii. Appendix B: Introduction

b. Identify readily available applicable databases or other information. *i. Appendix L and Appendix M*

c. Develop site characterization check list

i. Recommendations for Minimizing or Managing Injection-Induced Seismicity: Site Assessment Considerations

ii. Appendix B: Site Assessment Considerations for Evaluating Seismicity

d. Explore applicability of pressure transient testing and/or pressure monitoring techniques

i. Case Study Results

ii. Appendix D - Appendix H

3. Summary of lessons learned from case studies

i. Case Study Results: Common Characteristics, Observations, and Lessons Learned From Case Studies

4. Recommended measurement or monitoring techniques for higher risk areas

a. Approaches to address site assessment consideration

i. Appendix B and Appendix D

5. Applicability of conclusions to other well classes

Induced seismicity with respect to other well classes was discussed in the Introduction. The conclusions for the Class II disposal program may be applicable to other well classes; however, additional considerations may also be needed particularly for geothermal wells.

i. Appendix K: Subject Bibliography: Geothermal

6. Define if specific areas of research are needed
i. Research Needs

REPORT FINDINGS

The following major report findings are derived from the literature reviews, case study reviews, and the development of the decision model:

- The three key components behind injection-induced seismicity are (1) sufficient pressure buildup from disposal activities, (2) a Fault of Concern, and (3) a pathway for the increased pressure to communicate from the disposal well to the fault. Successful disposal occurs in areas with one or two characteristics present, but not all three.
- The UIC Director should take early prudent action to minimize the potential for injection-induced seismicity rather than requiring substantial proof of the causal relationship.
- The WG applied petroleum engineering techniques not identified in the injection-induced seismicity literature. These techniques have useful application for assessing flow path and fault presence. Basic petroleum engineering practices coupled with geology and geophysical information may provide a better assessment of well operational behavior in addition to improved understanding of reservoir and fault characteristics.
- A multidisciplinary approach is important for the evaluation of the key three components. Understanding the geologic characteristics and reservoir flow behavior of a site involves methodologies from petroleum engineering, geology and geophysics disciplines.
- The case studies were useful for identifying common characteristics of suspect wells and actions UIC Directors took through discretionary authority to manage and minimize seismic events in these areas.
- Additional research is needed to explore correlations between disposal well operational behavior and nearby earthquake events, taking into consideration all possible causal effects.
- Future research should consider a practical multidisciplinary approach and a holistic assessment addressing disposal well and reservoir behavior, geology and area seismicity.
- The decision model developed through this effort is based on a thought process derived from a combination of case studies, literature

reviews and understanding the conditions essential to cause seismicity. The WG selected a thought process versus a definitive framework to provide the Director with flexibility. The key questions of the decision model are:

- Have any seismicity concerns been identified in new or existing wells?
- Are there site considerations remaining following further review of data?
- Can a monitoring, operational or management approach be used to successfully address seismicity concerns?

Greater detail regarding these findings can be found in the respective report sections and associated appendices.

WG Project Team

Philip Dellinger, Lead	US EPA R6
Susie McKenzie, Technical Lead	US EPA R6
Nancy Dorsey, Technical Expert	US EPA R6
Ken Johnson, Technical Expert	US EPA R6
Rob Lawrence, R6 Policy Advisor	US EPA R6
Keara Moore	US EPA DC
William Bates	US EPA DC
Jill Dean	US EPA DC
Brian Graves	US EPA R6
Dave Rectenwald	US EPA R3
David Basinger	US EPA R9
George Robin	US EPA R9
Robert Smith	US EPA DC
Sarah Roberts	US EPA R8
Steve Platt	US EPA R3
Chuck Lowe	Ohio EPA
Tom Tomastik	Ohio Department of Natural Resources
Jim Milne	Colorado Oil and Gas Conservation Commission
Denise Onyskiw	Colorado Oil and Gas Conservation Commission, retired

Charles Lord	Oklahoma Corporation Commission
Vince Matthews	Colorado Geologic Survey, retired
Douglas Johnson	Railroad Commission of Texas
James A Peterson	West Virginia Department of Environmental Protection
Lawrence Bengal	Arkansas Oil and Gas Commission

ACKNOWLEDGMENTS

In addition to the members of the Working Group and National Technical Workgroup, the following technical experts participated in discussions or provided feedback on a working draft of this report.

Brian Stump, Southern Methodist University
Chris Hayward, Southern Methodist University
Scott Ausbrooks, Arkansas Geological Survey
Steve Horton, Center for Earthquake Research and Information, University of Memphis
Ernest Majer, Lawrence Berkeley National Laboratory
Norman Warpinski, Pinnacle
John Satterfield, formerly with Chesapeake Energy
Cliff Frohlich, Institute for Geophysics, University of Texas
David Dillon, National Academy of Science
Shah Kabir, Hess Energy
Bill Smith, National Academy of Science
Roy Van Arsdale, University of Memphis Justin Rubenstein, USGS

EPA's Office of Ground Water and Drinking Water, assisted by a contractor, submitted the report to the following experts for a final peer review.

Jeff Bull, Chesapeake Energy Corporation
Robin McGuire, Lettis Consultants International, Inc.
Craig Nicholson, University of California, Santa Barbara
Kris Nygaard, ExxonMobil
Heather Savage, Lamont-Doherty Earth Observatory, Columbia University
Ed Steele, Swift Worldwide Services

Project Review Team

The following EPA technical staff analyzed and incorporated changes based on the peer reviewers' recommendations and sent the results to the NTW for final review.

Philip Dellinger	US EPA R6
Nancy Dorsey	US EPA R6
Ken Johnson	US EPA R6
Rob Lawrence	US EPA R6
Arnold Bierschenk	US EPA R6
William Bates	US EPA DC
Bruce Kobelski	US EPA DC
Kurt Hildebrandt	US EPA R7

Glossary of Acronyms and Terms

Acronyms

AAPG	American Association of Petroleum Geologists
AGS	Arkansas Geological Survey
ANSS	USGS Advanced National Seismic System
AOGC	Arkansas Oil and Gas Commission
BHP	Bottomhole Pressure
CERI	Center for Earthquake Research and Information
Com	Cat Comprehensive catalog
DFW	Dallas-Fort Worth
EPA	U.S. Environmental Protection Agency
HF	Hydraulic Fracturing
IEA	International Energy Agency
M4.0	Magnitude earthquake event; for instance, M4.0 means magnitude 4.0
NCEER	Central and Eastern United States, CERI Earthquake database
NEIC	National Earthquake Information Center, U.S. Geological Survey
NTW	National Technical Workgroup
ODNR	Ohio Department of Natural Resources

OSN	Ohio Seismic Network
PDE	Preliminary Determination Earthquake, NEIC Earthquake database
RRC	Railroad Commission of Texas
SMU	Southern Methodist University
SPE	Society of Petroleum Engineers
SRA	Eastern, Central & Mountain States NEIC Earthquake database
UIC	Underground Injection Control
USDW	Underground Source of Drinking Water
USGS	U.S. Geological Survey
USHIS	Significant U.S. quakes, NEIC Earthquake database
WG	Injection-induced Seismicity Working Group
WVDEP	West Virginia Department of Environmental Protection Office of Oil and Gas

Terms

Catalog aka earthquake catalog from USGS online Earthquake Search of the NEIC PDE catalog of earthquakes. http://earthquake.usgs.gov /earthquakes/eqarchives/epic/

Class II injection wells are wells that inject fluids (1) which are brought to the surface in connection with conventional oil or natural gas production and may be commingled with waste waters from gas plants which are an integral part of production operations, unless those waters are classified as a hazardous waste at the time of injection, (2) for enhanced recovery of oil or natural gas; and (3) for storage of hydrocarbons which are liquid at standard temperature and pressure (40 CFR 146.5(b)).

Earthquake is a term used to describe both sudden slip on a fault, and the resulting ground shaking and radiated seismic energy caused by the slip, or by volcanic or magmatic activity, or other sudden stress changes in the earth (USGS). Earthquakes resulting from human activities will be called induced earthquakes in this report.

Epicenter is the point on the earth's surface vertically above the hypocenter (or focus) point in the crust where a seismic rupture begins. NEIC coordinates are given in the WGS84 reference frame. The position uncertainty of the hypocenter location varies from about 100 m horizontally and 300 m vertically for the best located events, those in the middle of densely spaced seismograph networks, to tens of kilometers for events in large parts of the United States.

Falloff test is a pressure transient test conducted by shutting an injection well in and observing the pressure decline at the well over a period of time.

Fault of Concern is a fault optimally oriented for movement and located in a critically stressed region. The fault is also of sufficient size, and possesses sufficient accumulated stress/strain, such that fault slip and movement has the potential to cause a significant earthquake. Fault may refer to a single fault or a zone of multiple faults and fractures. See also Geologic Stress Considerations; Appendix B: Site Assessment Considerations For Evaluating Seismicity; and Appendix M: State of Stress for more complete discussion.

Hypocenter, aka focus, is the 3-D location of the earthquake source, i.e., latitude, longitude, focal depth below ground.

Isopach is a contour map illustrating the variations of thickness of a defined stratum.

Magnitude is a number that characterizes the relative size of an earthquake at the hypocenter. Magnitude is based on the measurement of the maximum motion recorded by a seismograph or the energy released. Generally, damage is reported for magnitudes above 5.[21] Magnitude (M) will refer to the numbers reported by USGS or the NEIC, not separated between moment, body wave, or surface wave magnitudes.

Magnitude[22]	Earthquake Effects
2.5 or less	Usually not felt, but can be recorded by seismograph.
2.5 to 5.4	Often felt, but only causes minor damage.
5.5 to 6.0	Slight damage to buildings and other structures.
6.1 to 6.9	May cause a lot of damage in very populated areas.
7.0 to 7.9	Major earthquake. Serious damage.
8.0 or greater	Great earthquake. Can totally destroy communities near the epicenter.

Microseismicity has no formal definition, but generally is an earthquake with a magnitude less than 2. (*The Severity of an Earthquake*, USGS website: http://earthquake.usgs.gov/learn/topics/richter.php)

Step rate test is a pressure transient test that consists of a series of increasing injection rates as a series of rate steps and estimates the pressure necessary to fracture the formation.

Significant earthquakes/seismic events, for this report, are of a magnitude that can cause damage or potentially endanger underground sources of drinking water.

Static pressure, for this report, is the bottomhole pressure in the pore volume around the injection well measured in the wellbore at the end of a shut-in period that reaches stabilized conditions.

Tectonic is the rock structure and external forms resulting from the deformation of the earth's crust. (Dictionary of Geological Terms, 1976).

CITATIONS

Ake, J., K. Mahrer, D. O'Connell and L. Block, 2005, Deep-injection and closely monitored induced seismicity at Paradox Valley, Colorado: Bulletin Seismological Society, v. 95, no. 2, p. 664-683.

Ake, J., L. Block, and D. O'Connell 2002, What's shaking in bedrock? Paradox Valley deep-well injection program: Outcrop, v. 51, no. 4.

Alexander, S. S., R. Cakir, A. G. Doden, D. P. Gold and S. I. Root, (compilers), 2005, Basement depth and related geospatial database for Pennsylvania: Pennsylvania Geological Survey, 4th ser., Open-File General Geology Report 05-01.0, <http://www.dcnr.state.pa.us/topogeo /publications/pgspub/openfile/basementmap/index.htm>.
Accessed October 20, 2014

Ausbrooks, S.M. and E. Doerr, 2007, Enola swarm area of Faulkner County, Arkansas: GH-EQENOLA-002, Arkansas Geological Survey, 1 sheet.

Baranoski, M.T., 2002, in Structure contour map on the Precambrian unconformity surface in Ohio and related basement features, Ohio Department of Natural Resources, Division of Geological Survey Map PG-23.

Baranoski, M.T., 2013, in Structure contour map on the Precambrian unconformity surface in Ohio and related basement features, version 2; Ohio Department of Natural Resources, Division of Geological Survey Map PG-23, scale 500,000, 17 p.

Block, L., 2011, Paradox Valley deep disposal well and induced seismicity, Presented at National Academy of Sciences Meeting of the Committee on Induced Seismicity Potential in Energy Technologies: Dallas, Texas, Bureau of Reclamation, U.S. Department of the Interior.

British Columbia Oil and Gas Commission, 2012, Investigation of observed seismicity in the Horn River Basin: BC Oil and Gas Commission, 29 p.

Das, I., and M.D. Zoback, 2011, Long period long duration seismic events during hydraulic stimulation of a shale gas reservoir, Article #40761, Search and Discovery, American Association of Petroleum Geologists/Datapages, Inc.

Davis, S. D., and C. Frohlich, 1993, Did (or will) fluid injection cause earthquakes? Criteria for a rational assessment: Seismological Research Letters, v. 64, no. 3-4.

de Pater, C. J., and S. Baisch, 2011, Geomechanical study of Bowland Shale seismicity synthesis report: Cuadrilla Resources.

El Hariri, M., R. E. Abercrombie, C. A. Rowe and A. F. do Nascimento, 2010, Role of fluids in triggering earthquakes: Observations from reservoir induced seismicity in Brazil: Geophysical Journal International, v. 81, no. 3, p. 1566-1574.

Gerrish, H., and A. Nieto, 2003, Review of injection reservoir information in relation to earthquakes in Ashtabula, Ohio, 2nd International Symposium, Underground Injection Science and Technology: Symposium Abstracts: Berkeley, California, Lawrence Berkeley National Laboratory, p. 156.

Gidley, J. L., S. A. Holditch, D. E. Nierode, and R. W. Veatch, Jr., editors, 1990, Recent advances in hydraulic fracturing, SPE Monograph Series Volume 12, Society of Petroleum Engineers, p. 464.

Healy, J. H., W.W. Rubey, D.T. Griggs and C.B. Raleigh, 1968, Denver earthquakes: Science, v. 161, no. 3848, p. 1301-1310.

Holland, A. A., 2013, Optimal fault orientations within Oklahoma: Seismological Research Letters, v. 84, p. 876-890; doi:10.1785/0220 120153.

Holland, A., 2011, Examination of possibly induced seismicity from hydraulic fracturing in the Eola Field, Garvin County, Oklahoma, Oklahoma Geological Survey, Open-File Report OF1-2011.

Holland, A. A., C. R. Toth, and E. M. Baker, 2013, Probabilistic Seismic Hazard Assessment and Observed Ground Motions for the Arcadia, Oklahoma, Dam Site: Oklahoma Geological Survey Special Publication, SP2013-01, 63 p., (OGS website).

Holtkamp, S., B. Currie, and M. R. Brudzinski, 2013, A more complete catalog of the 2011 Youngstown, Ohio earthquake sequence from template matching reveals a strong correlation to pumping at a wastewater injection well, AAPG 2013 Annual Convention and Exhibition, Pittsburgh, Pennsylvania, May 19-22.

Horton, S. P., 2012, Disposal of hydrofracking-waste fluid by injection into subsurface aquifers triggers earthquake swarm in Central Arkansas with potential for damaging earthquakes: Seismological Research Letters, v. 83, p. 250-260.

Horton, S., and Ausbrooks, S., 2011, Earthquakes in central Arkansas triggered by fluid injection at Class 2 UIC wells, National Academy of Science Meeting of the Committee on Induced Seismicity Potential in Energy Technologies: Dallas, Texas.

Hsieh, P. A., and J. D. Bredehoeft, 1981, Reservoir analysis of the Denver earthquakes: a case of induced seismicity: Journal of Geophysical Research, v. 86, no. B2, p. 903-920.

Hunt, S. P., and C. P. Morelli, 2006, Cooper Basin HDR seismic hazard evaluation: Predictive modelling of local stress changes due to HFR geothermal energy operations in South Australia, *in* Adelaide, U. o., ed., South Australian Department of Primary Industries and Resources, Government of South Australia.

Janská, E. and L. Eisner, 2012, Ongoing seismicity in the Dallas-Fort Worth area: The Leading Edge, v. 31, p. 1462-1468.

Johnson, D., 2011, Regulatory response to induced seismicity in Texas, Presented at National Academy of Sciences Meeting of the Committee on Induced Seismicity Potential in Energy Technologies: Dallas, Texas, Texas Railroad Commission.

Kanamori, H. and E. Hauksson, 1992, A slow earthquake in the Santa Maria Basin, California: Bulletin of the Seismological Society of America, v. 82, p. 2087-2096.

Kim, W-Y, 2013, Induced seismicity associated with fluid injection into a deep well in Youngstown, Ohio, Journal of Geophysical Research: Solid Earth, v. 118, p. 3506- 3518.Lee, J., J. B. Rollin, and J. P. Spivey, 2003, Pressure transient testing, SPE Textbook Series, Society of Petroleum Engineers.

Kim, W-Y., , J. Armbruster, M. Hansen, L. Wickstrom, C. Grope, J. Dick and W. Leith, 2012, Youngstown Earthquake on 24 December 2011 and 31 December 2011; Appendix 2-

Lamont-Doherty Ohio UIC Shutdown, Ohio Department of Natural Resources, 5 p.

Klose, C., 2013, Mechanical and statistical evidence of the causality of human-made mass shifts on the Earth's upper crust and the occurrence of earthquakes: Journal of Seismology, v. 17, n. 1, p. 109-135; doi:10.1007/s10950-012-9321-8.

Mahrer, K., J. Ake, L. Block, D. O'Connell and J. Bundy, 2005, Injecting brine and inducing seismicity at the world's deepest injection well, Paradox

Valley, Southwest Colorado: Developments in Water Science, v. 52, p. 361-375.

Majer, E. L., R. Baria, M. Stark, S. Oates, J. Bommer, B. Smith and H. Asanuma, 2007, Induced seismicity associated with enhanced geothermal systems: Geothermics, v. 36, p. 185- 222.

Majer, E., J. Nelson, A. Robertson-Tait, J. Savy, and I. Wong, 2011, Protocol for addressing induced seismicity associated with enhanced geothermal systems, <http://www1.eere.energy.> Accessed November 22, 2011.

Maxwell, S. C., D. Cho, T. Pope, M. Jones, C. Cipolla, M. Mack, F. Henery, M. Norton and J. Leonard et al., 2011, Enhanced reservoir characterization using hydraulic fracturing microseismicity: SPE 140449-MS, SPE Hydraulic Fracturing Technology Conference, 24- 26 January 2011, The Woodlands, Texas, USA, p. 457-467.

National Research Council, 2013, Induced seismicity potential in energy technologies, Washington, D.C., The National Academies Press, 262 p., <http://www.nap.edu/catalog.php?record id=13355>.

Nicholson, C., and R. L. Wesson, 1990, Earthquake hazard associated with deep well injection— A report to the U.S. Environmental Protection Agency, *in* U.S. Geological Survey Bulletin 1951, 74 pp. plus plate.

Nicholson, C., and R. L. Wesson, 1992, Triggered earthquakes and deep well activities: Pure and Applied Geophysics, v. 139, no. 3-4, p. 561-568.

Ohio Department of Natural Resources, 2012, preliminary report on the Northstar 1 Class II injection well and the seismic events in the Youngstown, Ohio area: Ohio Department of Natural Resources, 24 p. plus figures, <http://www.ohiodnr.com/homepage/NewsReleases/tabid/18276/EntryId/2711/Ohios-New-Rules-for-Brine-Disposal-Among-Nations-Toughest.aspx>

Perkins, T. K. and J. A. Gonzalez, 1984, Changes in Earth stresses around a wellbore caused by radially symmetrical pressure and temperature gradients: SPEJ April 1984, pp 129-140.

Phillips, W. S., J. T. Rutledge, L. S. House and M. C. Fehler, 2002, Induced microearthquake patterns in hydrocarbon and geothermal reservoirs: Six case studies: Pure and Applied Geophysics, v. 159, no. 1-3, p. 345-369.

Raleigh, C. B., 1972, Earthquakes and fluid injection: Experiment in earthquake control at Rangely, Colorado, AAPG Memoir 18: AAPG Special Volumes, American Association of Petroleum Geologists.

Raleigh, C. B., J. H. Healy, and J. D. Bredehoeft, 1976, An experiment in earthquake control at Rangely, Colorado: Science, v. 191, no. 4233, p. 1230-1237.

Seeber, L., and J. Armbruster, 1993, Natural and induced seismicity in the Lake Erie-Lake Ontario region: reactivation of ancient faults with little neotectonic displacement: Géographie physique et Quaternaire, v. 47, n. 3, p. 363-378.

Seeber, L., and J. Armbruster, 2004, A fluid-injection-triggered earthquake sequence in Ashtabula, Ohio: Implications for seismogenesis in stable continental regions: Bulletin of the Seismological Society of America, v. 94 n. 1, p. 76-87.

Suckale, J., 2009, Induced seismicity in hydrocarbon fields: Advances in Geophysics, Academic Press, p. 55-106.

Suckale, J., 2010, Moderate-to-large seismicity induced by hydrocarbon production: The Leading Edge, v. 29, no. 3, p. 310-319.

Tingay, M. R. P., B. Müller, J. Reinecker and O. Heidbach, 2006, State and origin of the present-day stress field in sedimentary basins: New results from the stress map project, ARMA/USRMS 06-1049, Golden Rocks 2006, The 41st U.S. Symposium on Rock Mechanics (USRMS), Golden, Colorado.

Warpinski, N. R., M. J. Mayerhofer, M. C. Vincent, C. L. Cipolla and E. P. Lolon, 2008, Stimulating unconventional reservoirs: maximizing network growth while optimizing fracture conductivity: J Can Pet Technol 48 (10): 39-51; doi:10.2118/114173-PA; SPE 114173.

Warpinski, N., 2009, Microseismic monitoring: inside and out: Journal of Petroleum Technology, v. 61, no. 11, p. 80-85.

Warpinski, N., J. Du, and U. Zimmer, 2012, Measurements of hydraulic-fracture-induced seismicity in gas shales: SPE 151597 presented at the Society of Petroleum Engineers Hydraulic Fracturing Technology Conference, The Woodlands, Texas, 6-8 February.

End Notes

[1] Information on earthquake terms is included under Glossary Terms In the full report or at http://earthquake.usgs.gov/learn/glossary/ for general earthquake terms

[2] For the purposes of this report, the Induced Seismicity Working Group considers "significant" seismic events to be those of a magnitude that could potentially cause damage to or endanger underground sources of drinking water.

[3] National Research Council. Chapter 3, Table 3.4, page 104, and Chapter 7, injection wells for the disposal of water associated with energy extraction finding no. 1, *in* Induced seismicity potential in energy technologies. (Washington, DC: The National Academies Press, 2013), 104 and 171–172.

[4] R.L. Wesson and C. Nicholson, Earthquake hazard associated with deep well injection: A report to the U.S. Environmental Protection Agency, U.S. Geological Survey Open-file Report 87-331, 1987, 108 pp.

[5] 40 CFR §144.12(b); 40 CFR §144.52(a)(9) or (b)(1); or appropriate section of 40 CFR Part 147

[6] A Fault of Concern is a fault optimally oriented for movement and located in a critically stressed region. The fault is also of sufficient size and possesses sufficient accumulated stress / strain, such that fault slip and movement has the potential to cause a significant earthquake. Fault may refer to a single fault or a zone of multiple faults and fractures. See also Geologic Stress Considerations later in this document; Appendix B: Site Assessment Considerations for Evaluating Seismicity; and Appendix M: State of Stress for more complete discussion.

[7] Seismic events resulting from human activities are referred to as induced, for this report.

[8] Magnitude will refer to the values reported by the USGS Advanced National Seismic System catalog.

[9] Chapter 3, Table 3.4, page 104, and Chapter 7, Injection Wells for the Disposal of Water Associated with Energy Extraction Finding No. 1, pages 171-172; "Induced Seismicity Potential in Energy Technologies," 2013 NAS Publication.

[10] For the purposes of this report, "significant" seismic events are of a magnitude that has the potential to cause damage or endanger underground sources of drinking water or cause infrastructure damage.

[11] Earthquake terms are included under *Glossary Terms* later in this report or http://earthquake.usgs.gov/learn/glossary/ for general earthquake terms.

[12] A Fault of Concern is a fault optimally oriented for movement and located in a critically stressed region. The fault is also of sufficient size, and possesses sufficient accumulated stress/strain, such that fault slip and movement has the potential to cause a significant earthquake. Fault may refer to a single fault or a zone of multiple faults and fractures.

[13] 40 CFR §146.62(b)(1) and §146.68(f) for Class I hazardous; §146.82(a)(3)(v) for Class VI geologic sequestration

[14] 40 CFR §144.12(b); 40 CFR §144.52(a)(9) or (b)(1); or appropriate section of 40 CFR Part 147

[15] An earlier draft version (available only in EPA files) was assumed to have been replaced by the final publication.

[16] Microseismic and small seismic events may occur but go undetected or unfelt and pose no significant risk to human health or USDWs.

[17] RRC of TX website: http://www.rrc.state

[18] http://codes.ohio.gov/oac/1501%3A9-3-06

[19] http://codes.ohio.gov/oac/1501%3A9-3-07

[20] Many of these parameters may be requested, but not required.

[21] Building damage was reported following 2011 earthquakes near Trinidad, Colorado (5.3); near Greenbrier, Arkansas (4.7); and the Soultz, France, project (2.9).

[22] Michigan Tech, <http://www.geo.mtu.edu/UPSeis/magnitude>, Accessed 11/10/14.

In: Deep-Well Injections and Induced Seismicity ISBN: 978-1-63482-557-3
Editor: James B. Fernandez

Chapter 3

EARTHQUAKES: RISK, DETECTION, WARNING, AND RESEARCH*

Peter Folger

SUMMARY

Portions of all 50 states and the District of Columbia are vulnerable to earthquake hazards, although risks vary greatly across the country and within individual states. Seismic hazards are greatest in the western United States, particularly in California, Washington, Oregon, and Alaska and Hawaii. California has more citizens and infrastructure at risk than any other state because of the state's frequent seismic activity combined with its large population and developed infrastructure.

The United States faces the possibility of large economic losses from earthquake-damaged buildings and infrastructure. The Federal Emergency Management Agency has estimated that earthquakes cost the United States, on average, over $5 billion per year. California, Oregon, and Washington account for nearly $4.1 billion (77%) of the U.S. total estimated average annualized loss. California alone accounts for most of the estimated annualized earthquake losses for the nation.

A single large earthquake, however, can cause far more damage than the average annual estimate. The 1994 Northridge (CA) earthquake caused as much as $26 billion (in 2005 dollars) in damage and was one of the

* This is an edited, reformatted and augmented version of a Congressional Research Service publication RL33861, prepared for Members and Committees of Congress dated July 18, 2013.

costliest natural disasters to strike the United States. One study of the damage caused by a hypothetical magnitude 7.8 earthquake along the San Andreas Fault in southern California projected as many as 1,800 fatalities and more than $200 billion in economic losses.

Unlike other natural hazards, such as hurricanes, where predicting the location and timing of landfall is becoming increasingly accurate, the scientific understanding of earthquakes does not yet allow for precise earthquake prediction. Instead, notification and warning typically involve communicating the location and magnitude of an earthquake as soon as possible after the event to emergency response providers and others who need the information.

A precise relationship between earthquake mitigation measures, federal earthquake-related activities such as earthquake research, and reduced losses from an actual earthquake may never be possible. However, as more accurate seismic hazard maps evolve, and as understanding of the relationship between ground motion and building safety improves, trends denoting the effectiveness of mitigation strategies and earthquake research and other activities may emerge more clearly. Without an ability to precisely predict earthquakes, Congress is likely to face an ongoing challenge in determining the most effective federal approach to increasing the nation's resilience to low-probability but high-impact major earthquakes.

INTRODUCTION

Close to 75 million people in 39 states face some risk from earthquakes. Earthquake hazards are greatest in the western United States, particularly in California, but also in Alaska, Washington, Oregon, and Hawaii. Earthquake hazards are also prominent in the Rocky Mountain region and the New Madrid Seismic Zone (a portion of the central United States), as well as in portions of the eastern seaboard, particularly South Carolina. Under the National Earthquake Hazards Reduction Program (NEHRP), the federal government supports efforts to assess and monitor earthquake hazards and risk in the United States.[1] Given the potentially huge costs associated with a large, damaging earthquake in the United States, an ongoing issue for Congress is whether the federally supported earthquake programs are appropriate for the earthquake risk. This report discusses:

- earthquake hazards and risk in the United States,
- federal programs that support earthquake monitoring,

- the U.S. capability to detect earthquakes and issue notifications and warnings, and
- federally supported research to improve the fundamental scientific understanding of earthquakes with a goal of reducing U.S. vulnerability.

EARTHQUAKE HAZARDS AND RISK

Portions of all 50 states and the District of Columbia are vulnerable to earthquake hazards, although risks vary greatly across the country and within individual states. (See, for example, the box below describing the August 23, 2011, magnitude 5.8 earthquake in Virginia.) Seismic hazards are greatest in the western United States, particularly in California, Washington, Oregon, and Alaska and Hawaii. Alaska is the most earthquake-prone state, experiencing a magnitude 7 earthquake almost every year and a magnitude 8 earthquake every 14 years on average. (See box below for a description of earthquake magnitude.) Because of its low population and infrastructure density, Alaska has a relatively low risk for large economic losses from an earthquake. In contrast, California has more citizens and infrastructure at risk than any other state because of the state's frequent seismic activity combined with its large population.

United States National Seismic Hazard Map

Figure 1 shows where earthquakes are likely to occur in the United States and how severe the earthquake magnitude and resulting ground shaking are likely to be. The map in Figure 1 depicts the potential shaking hazard from future earthquakes. It is based on the frequency at which earthquakes occur in different areas and how far the strong shaking extends from the source of the earthquake. In Figure 1, the hazard levels indicate the potential ground motion—expressed as a percentage of the acceleration due to gravity (g). In a sense, the map shows the likelihood of where earthquakes could occur, and where the strongest shaking could take place.

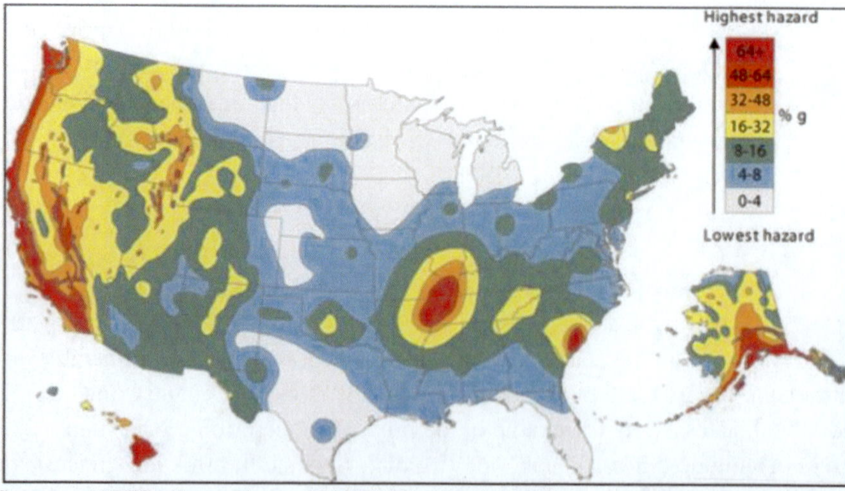

Source: USGS Fact Sheet 2008-3018 (April 2008), at http://pubs.usgs.gov/fs/2008/3018/pdf/FS08-3018_508.pdf. Modified by CRS.

Note: The bar in the upper right shows the potential ground motion—expressed as a percentage of the acceleration due to gravity (g)—with up to a 1 in 50 chance of being exceeded over a 50-year period.

Figure 1. Earthquake Hazard in the United States.

Earthquake Magnitude and Intensity

Earthquake magnitude is a number that characterizes the relative size of an earthquake. It was historically reported using the Richter scale (magnitudes in this report are generally consistent with the Richter scale). Richter magnitude is calculated from the strongest seismic wave recorded from the earthquake, and is based on a logarithmic (base 10) scale: for each whole number increase in the Richter scale, the ground motion increases by 10 times. The amount of energy released per whole number increase, however, goes up by a factor of 32. The moment magnitude scale is another expression of earthquake size, or energy released during an earthquake, that roughly corresponds to the Richter magnitude and is used by most seismologists because it more accurately describes the size of very large earthquakes. Sometimes earthquakes will be reported using qualitative terms, such as Great or Moderate. Generally, these terms refer to magnitudes as follows: Great (M>8); Major (M>7); Strong (M>6); Moderate (M>5); Light (M>4); Minor (M>3); and Micro (M<3).

Intensity is a measure of how much shaking occurred at a site based on observations and amount of damage. Intensity is usually reported on the Modified Mercalli Intensity Scale as a Roman numeral ranging from I (not felt) to XII (total destruction). The intensity of an earthquake depends on where the earthquake occurs, how it is felt by people, and the damage it causes. The lower numbers of the Modified Mercalli Intensity Scale generally refer to how the earthquake is felt by people, and the higher numbers are based on observed structural damage.

Modified Mercalli intensities that are typically observed at locations near the epicenters of earthquakes of different magnitudes are as follows:

Magnitude 1.0-3.0 Modified Mercalli Intensity I
Magnitude 3.0-3.9 Modified Mercalli Intensity II-III
Magnitude 4.0-4.9 Modified Mercalli Intensity IV-V
Magnitude 5.0-5.9 Modified Mercalli Intensity VI-VII
Magnitude 6.0-6.9 Modified Mercalli Intensity VII-IX
Magnitude 7.0+ Modified Mercalli Intensity VIII or higher

Source: USGS FAQs, at http://earthquake.usgs.gov/learn/faq/; and Magnitude/ Intensity Comparison, at http://earthquake.usgs.gov/learn/topics/ mag_vs_ int.php.

Figure 1 also shows relatively high earthquake hazard in the Rocky Mountain region, portions of the eastern seaboard—particularly South Carolina—and a part of the central United States known as the New Madrid Seismic Zone (see "The New Madrid Seismic Zone" below). Other portions of the eastern and northeastern United States are also vulnerable to moderate seismic hazard. According to the USGS, 75 million people in 39 states are subject to "significant risk."[2]

August 23, 2011, Magnitude 5.8 Earthquake in Virginia

At 1:51 p.m. on August 23, 2011, an unusually large magnitude 5.8 earthquake struck near the town of Mineral, VA, about 38 miles northwest of Richmond and 84 miles southwest of Washington, DC. According to the U.S. Geological Survey, the relatively shallow earthquake (focal depth of 3.7 miles) occurred within what is referred to as the "Central Virginia Seismic Zone." Small to moderate earthquakes have occurred within the Central Virginia Seismic Zone since at least the 18th century, although the earthquakes have not been directly associated with any mapped faults. The fault responsible for the August 23 earthquake is also unmapped, but more detailed seismological investigation in the wake of the earthquake may identify the fault segment. Earthquakes of a similar magnitude typically involve slippage along fault segments 3 to 9 miles long, according to the USGS. (The USGS thinks that the earthquake occurred on a "reverse" fault, where one side of a steeply dipping fault moves up and over the other side.) The largest earthquake that occurred previously in the Central Virginia Seismic Zone occurred in 1875 and was approximately magnitude 4.8, based on the area experiencing shaking (modern seismological instruments were not available in 1875).

The August 23, 2011, earthquake demonstrates how seismic waves generated in the eastern part of the country are felt over a much wider radius than would occur with waves from a similar earthquake in California. The USGS observes that an earthquake east of the Rocky Mountains may be felt over an area 10 times as large as that from a California earthquake of the same magnitude. The geology underlying California and portions of other western states generally has more faults than in the nation's interior and on the East Coast. Seismic waves are not transmitted across the faults as efficiently as they are in the older, denser, and colder rocks that occur in the east. According to the USGS "Did You Feel It?" website, shaking from the August 23, 2011, earthquake was felt as far south as South Carolina and into Georgia, and as far north as southern Ontario.

Source: USGS, 2011 August 23 Earthquake Summary, http://earthquake.usgs.gov/earthquakes/recenteqsww/Quakes/ se082311a.html#summary.

2008 Update to the National Seismic Hazard Map

In 2008, the USGS released National Seismic Hazards Maps that updated the version published in 2002.[3] Compared to the 2002 version, the new maps indicate lower ground motions (by 10% to 25%) for the central and eastern United States, based on modifications to the ground-motion models used for earthquakes. The new maps indicate that estimates of ground motion for the western United States are as much as 30% lower for certain types of ground motion, called long-period seismic waves, which affect taller, multi-story buildings. Ground motion that affects shorter buildings of a few stories, called short-period seismic waves, is roughly similar to the 2002 maps. The new maps show higher estimates for ground motion for western Oregon and

Washington compared to the 2002 maps, due to new ground motion models for the offshore Cascadia subduction zone. In formulating the 2008 maps, the USGS gave more weight to the probability of a catastrophic magnitude 9 earthquake occurring along the Cascadia subduction zone. The Cascadia subduction zone fault ruptures, on average, every 500 years, and has the potential to generate destructive earthquakes and tsunamis along the coasts of Washington, Oregon, and northern California.

Earthquake Forecast for California

According to a report released on April 14, 2008, California has a 99% chance of experiencing a magnitude 6.7 or larger earthquake in the next 30 years.4 The likelihood of an even larger earthquake, magnitude 7.5 or greater, is 46%, and such an earthquake would likely occur in the southern part of the state. The fault with the highest probability of generating at least one earthquake of magnitude 6.7 or greater over the next 30 years is the San Andreas in southern California (59% probability); for northern California it is the Hayward-Rodgers Creek fault (31%). The earthquake forecasts are not predictions (i.e., they do not give a specific date or time), but represent probabilities over a given time period. In addition, the probabilities have variability associated with them. The earthquake forecasts are known as the "Uniform California Earthquake Rupture Forecast (UCERF)" and are produced by a working group composed of the USGS, the California Geological Survey, and the Southern California Earthquake Center.

Earthquake Fatalities

Since 2000, only two deaths directly caused by earthquakes have occurred in the United States, both associated with falling debris in Paso Robles (CA) during the December 22, 2003, San Simeon earthquake of magnitude 6.5. In contrast, earthquakes have been directly or indirectly responsible for more than 685,000 fatalities in other countries since 2000.[5] Approximately 65% of those estimated deaths resulted from the December 2004 Indonesian earthquake (and resulting tsunami) of magnitude 9.1, and the January 2010 magnitude 7.0 earthquake in Haiti.

About 98% of earthquakes detected each year by the National Earthquake Information Center (NEIC) are smaller than magnitude 5.0 (light earthquakes); only 63 earthquakes exceeded magnitude 6.0 (strong earthquakes) for the 10-year period (about 0.2% of the total earthquakes detected), for an average of about six earthquakes per year of at least 6.0 magnitude. Larger earthquakes, although infrequent, cause the most damage and are responsible for most earthquake-related deaths. The great San Francisco earthquake of 1906 claimed an estimated 3,000 lives, as a result of both the earthquake and subsequent fires. Over the past 100 years, relatively few Americans have died as a result of earthquakes, compared to citizens in some other countries.[6] Since 1970, three strong earthquakes (greater than magnitude 6) in the United States were responsible for 188 of the 212 total earthquake-related fatalities. (See *Table 1.*)

Table 1. Earthquakes Responsible for Most U.S. Fatalities Since 1970

Date		Magnitude	Deaths
February 9, 1971	San Fernando Valley, CA	6.6	65
October 18, 1989	Loma Prieta, CA	6.9	63
January 17, 1994	Northridge, CA	6.7	60

Source: USGS, http://earthquake.usgs.gov/earthquakes/states/us_deaths.php.
Note: Other sources report different numbers of fatalities associated with the Northridge earthquake.

Estimating Potential Losses from Earthquakes

Estimating the seismic hazard for a region—as in *Figure 1*—is a first step in assessing risk. As a second step, shaking hazards maps are often combined with other data, such as the strength of existing buildings, to estimate possible damage in an area due to an earthquake. A third step in estimating potential losses would be in assigning value to the infrastructure at risk from earthquake damage. The combination of seismic risk, population, and vulnerable infrastructure can help improve the understanding of which urban areas across the United States face risks from earthquake hazards that may not be immediately obvious from the probability maps of shaking hazards alone, and what potential economic costs may be at stake.

The 1994 Northridge earthquake was the nation's most damaging earthquake in the past 100 years, preceded five years earlier by the second-most costly earthquake—Loma Prieta. Comparing losses between different earthquakes, and between earthquakes and other disasters such as hurricanes, can be difficult because of the different ways losses are calculated. Calculations may include a combination of insured losses, uninsured losses, and estimates of lost economic activity.

The United States faces potentially large total losses due to earthquake-caused damage to buildings and infrastructure and lost economic activity. As urban development continues in earthquake-prone regions in the United States, concerns are increasing about the exposure of the built environment, including utilities and transportation systems, to potential earthquake damage.[7] One estimate of economic loss from a severe earthquake in the Los Angeles area is over $500 billion.[8] Another estimate of economic loss from a hypothetical 6.5 magnitude earthquake along the heavily populated central New Jersey-Philadelphia corridor would be even higher— approximately $900 billion. The seismic hazard in the New Jersey-Philadelphia regions, however, is much lower than in the Los Angeles area, as shown in *Table 2.*

Table 2. U.S. Metropolitan Areas with Estimated Annualized Earthquake Losses of More Than $10 Million in $millions

Rank	Metro area	AEL	Rank	Metro area	AEL
1	Los Angeles-Long Beach-Santa Ana, CA	$1,31 2	23	Reno-Sparks, NV	$29
2	San Francisco-Oakland-Fremont, CA	$781	24	Charleston-North Charleston, SC	$22
3	Riverside-San Bernadino-Ontario, CA	$397	25	Columbia, SC	$22
4	San Jose-Sunnyvale-Santa Clara, CA	$277	26	Stockton, CA	$21
5	Seattle-Tacoma, WA	$244	27	Atlanta-Sandy Springs-Marietta, GA	$19
6	San Diego-Carlsbad-San Marcos, CA	$155	28	Bremerton-Silverdale, WA	$18
7	Portland-Vancouver-Carlsbad, OR	$137	29	Ogden-Clearfield, UT	$18
8	Oxnard-Thousand Oaks-Ventura, CA	$111	30	Salem, OR	$17
9	Santa Rosa-Petaluma, CA	$69	31	Eugene-Springfield, OR	$17
10	St. Louis, MO-IL	$59	32	Napa, CA	$16
11	Salt Lake City, UT	$52	33	San Luis Obispo-Paso Robles, CA	$16
12	Sacramento-Arden-Arcade-Roseville, CA	$52	34	Nashville-Davidson-Murfreesboro, TN	$15
13	Vallejo-Fairfield, CA	$40	35	Albuquerque, NM	$15
14	Memphis, TN	$38	36	Olympia, WA	$14
15	Santa Cruz-Watsonville, CA	$36	37	Modesto, CA	$13
16	Anchorage, AK	$35	38	Fresno, CA	$13

Table 2. (Continued)

Rank	Metro area	AEL	Rank	Metro area	AEL
17	Santa Barbara-Santa Maria-Goleta, CA	$34	39	Evansville, IN-KY	$12
18	Las Vegas-Paradise, NV	$33	40	Birmingham-Hoover, AL	$11
19	Honolulu, HI	$32	41	El Centro, CA	$11
20	Bakersfield, CA	$30	42	Little Rock-North Little Rock, AR	$11
21	New York-Northern New Jersey- Long Island, NY	$30	43	Provo-Orem, UT	$10
22	Salinas, CA	$29			

Source FEMA Publication 366, HAZUS MH Estimated Annualized Earthquake Losses for the United States (April 2008). Annualized earthquake losses (AEL) calculated in 2005 dollars.

Another approach to estimating potential losses is to "normalize" the damage estimates from past earthquakes by adjusting for inflation, increases in wealth, and changes in population. For example, adjusting the 1906 San Francisco earthquake and subsequent fire using 2005 dollars results in between $39 billion and $328 billion in losses, depending on assumptions and earthquake mitigation measures if that earthquake happened today.[9]

Some studies and techniques combine seismic risk with the value of the building inventory[10] and income losses (e.g., business interruption, wage, and rental income losses) in cities, counties, or regions across the country to provide estimations of economic losses from earthquakes. An April 2008 report from the Federal Emergency Management Agency (FEMA) calculated that the average *annualized* loss from earthquakes nationwide is $5.3 billion, with California, Oregon, and Washington accounting for nearly $4.1 billion (77%) of the U.S. total estimated average annualized loss.[11] *Table 2* shows metropolitan areas with estimated average annualized U.S. earthquake losses over $10 million.

Annualized earthquake loss (AEL) addresses two components of seismic risk: the probability of ground motion and the consequences of ground motion. It enables comparison between different regions with different seismic hazards and different building construction types and quality. For example, earthquake hazard is higher in the Los Angeles area than in Memphis, but the general building stock in Los Angeles is more resistant to the effects of earthquakes. The AEL annualizes the expected losses by averaging them by year.

A single large earthquake can cause far more damage than the average annual estimate. Annualized estimates, however, help provide comparisons of infrequent, high-impact events like damaging earthquakes with more frequently occurring hazards like floods, hurricanes, or other types of severe weather. The annualized earthquake loss values shown in *Table 2* represent future estimates, and are calculated by multiplying losses from potential future ground motions by their respective frequencies of occurrence, and then summing these values.[12]

Table 2 shows that annualized earthquake losses in the New York-Northern New Jersey-Long Island metropolitan area are $30 million (ranked 21 out of 43 metropolitan areas with losses greater than $1 million per year), even though no destructive earthquakes have struck that area for generations.[13] This area has a relatively low seismic hazard, but also has an extensive infrastructure and is densely populated. That combination of seismic risk, extensive infrastructure, and dense population produces a significant risk to people and structures, according to some estimates.[14]

A Decrease in Estimated Loss?

In its 2008 publication estimating potential earthquake losses, FEMA noted that the $5.3 billion in annualized earthquake loss nationwide was 21% higher than the $4.4 billion calculated in FEMA's previous report, published in February 2001.[15] However, the 2001 report calculated losses using 1994 dollars, and when adjusted to reflect 2005 dollars the earlier estimate increased to $5.6 billion, indicating a small decrease in nationwide annualized earthquake loss potential since the 2001 report was published. According to FEMA, this loss occurred even though the national building inventory increased by 50% over this same period.

What factors led to a decreased estimate in potential loss despite growth in building inventory? According to FEMA, two primary factors were responsible: (1) a slight decrease in estimated earthquake hazard in the western United States (namely California) except for some parts of Washington and Utah, and (2) a change in the distribution of building inventory in California, with a 17% increase in wood frame buildings and a reduction in the amount of masonry (-6%), steel (-5.8%), and concrete (-3%) buildings in the state.[16] Wood frame buildings are less vulnerable to earthquake damage, generally, compared to other construction types. Because California accounts for 66% of the overall nationwide annualized earthquake loss, a 17% increase in wood frame buildings had a proportionally large effect. In fact, FEMA attributed 78% of the loss reduction between 2001 and 2008 to the change in building inventory distribution, and 22% to the decrease in earthquake hazard for California.[17]

The New Madrid Seismic Zone

The New Madrid Seismic Zone in the central United States is vulnerable to large but infrequent earthquakes. A series of large (magnitude greater than 7.0) earthquakes struck the Mississippi Valley over the winter of 1811-1812, centered close to the town of New Madrid, MO. Some of the tremors were felt as far away as Charleston, SC, and Washington, DC. The mechanism for the earthquakes in the New Madrid zone is poorly understood,[18] and no earthquakes of comparable magnitude have occurred in the area since these events.

Estimating earthquake damage is not an exact science and depends on many factors. As described above, these are primarily the probability of ground motion occurring in a particular area (see *Figure 1*), and the consequences of that ground motion, which are largely a function of building construction type and quality, and of the level of ground motion and shaking during the actual event. Such factors contribute to the difficulty of making a reasonable damage estimate for a low-frequency, high-impact event in the New Madrid region based on the probability of an earthquake of similar magnitude occurring. This uncertainty has implications for policy decisions to ameliorate risk, such as setting building codes, and for designing and building structures to withstand a level of shaking commensurate with the risk. Presumably, the same seismic hazard should lead to similar building codes in urban areas (e.g., in *Figure 1*, compare the seismic hazard for the New Madrid area with portions of California).

Some researchers have questioned whether erring on the side of caution in the New Madrid Seismic Zone is justified.[19] These researchers challenge whether the benefits of building structures to conform with the earthquake probability estimates merit the costs, in light of the uncertainty in making those probability estimates.[20] These analyses may call into question whether the probability of ground motion estimates for the New Madrid Seismic Zone (the bullseye-shaped area shown in *Figure 1* that includes parts of Arkansas, Illinois, Tennessee, and Missouri), and other regions of the country that experience infrequent earthquakes, are too high.[21] A contributing factor to the uncertainty in estimating the earthquake hazard in the New Madrid Seismic Zone is the small amount of ground motion measured across the major faults, compared to much faster motions measured across major faults in California.[22] Typically, seismologists estimate the stress that builds up on a fault by measuring ground motion across the fault: the faster the motion, the more quickly the stress builds up. The buildup of stress may be ultimately released in an earthquake during which the rocks on one side of the fault move relative to the other side. Generally, for fast-moving faults such as the San Andreas Fault, the period of earthquake recurrence is short compared to faults where the ground motion is relatively slow.

Yet despite the uncertainty raised by some researchers because of the apparent lack of much ground motion, the USGS attributes a seismic hazard to areas of the New Madrid Seismic Zone comparable to the most seismically active portions of California (see *Figure 1*), where earthquakes are much more frequent, and the mechanisms for generating earthquakes are better understood. The seismic hazard in the New Madrid Seismic Zone is based largely on its past history, namely the three major earthquakes that struck the region 200 years ago. The current lack of much ground motion is a confusing factor for scientists trying to understand the New Madrid Seismic Zone.

Earthquakes in Haiti, Chile, and Japan—Some Comparisons

The magnitude 8.8 earthquake that struck Chile on February 27, 2010, was over 60 times larger than the magnitude 7.0 earthquake that destroyed Port-au-Prince, Haiti, less than two months earlier. Yet the number of deaths and the amount of damage in Haiti far exceeded damage and fatalities in Chile. The Chile earthquake occurred offshore, and was deeper and farther away from major cities than the Haiti earthquake; in addition, the infrastructure in Chile—buildings, highways, bridges—appears to have been built to withstand earthquake shaking far better than similar infrastructure in Haiti. Japan's magnitude 9.0 earthquake on March 11, 2011, was even larger and more destructive than the Chile earthquake, but a large portion of the damage was caused by a powerful tsunami. The three countries faced significant seismic hazards, although the hazards facing Chile and Japan were arguably better known, because Chile experienced a great (magnitude 9.5) earthquake in 1960[23] and Japan experienced a very damaging earthquake in Kobe in 1995 and has a long history of seismic activity. By contrast, Haiti had last experienced a large earthquake in 1860 (earthquakes in 1751 and 1770 destroyed Port-au-Prince; the 1860 earthquake struck farther west). In addition to the seismic *hazard*, which is a consequence of geology and plate tectonics,

Haiti's vulnerability to earthquake shaking appears to have exceeded Chile's. Japan's dense population and infrastructure, in particular the nuclear power reactors located on the northeast coastline close to the epicenter, increased its vulnerability to the March 11, 2011, earthquake and tsunami. However, Haiti was at greater *risk* of fatalities—from the earthquake and resulting damage to buildings—than Chile or Japan, even though Japan's 2011 earthquake was approximately 100 times larger than the Haiti earthquake.

January 12, 2010, Magnitude 7.0 Earthquake in Haiti

On Tuesday, January 12, 2010, a magnitude 7.0 earthquake struck Haiti at 4:53 p.m. The epicenter was located approximately 15 miles west-southwest of Port-au-Prince, and the earthquake occurred at a depth of about 8 miles, according to the USGS.[24] The relatively shallow earthquake, and its close proximity to the capital city, exposed millions of Haitians to severe-toviolent ground shaking. The earthquake occurred along the Enriquillo-Plantain Garden fault system, a major east-west trending strike-slip fault system that lies between the Caribbean tectonic plate and the North American tectonic plate; the Caribbean plate actively moves against the North American plate and shear stresses are created at the boundary. At a strike-slip fault, the rocks move past each other horizontally along the fault line (in contrast to a thrust fault, where rocks on one side of the fault move on top of the rocks on the other side). Other examples of strike-slip faults are the San Andreas fault in California and the Red River fault in China.

The January 12, 2010, earthquake caused widespread damage in the Port-au-Prince area, causing approximately 223,000 deaths and 300,000 injuries.[25] Also, a series of aftershocks followed the main earthquake. There were 14 aftershocks greater than magnitude 5, and 36 greater than magnitude 4, within the first day following the magnitude 7.0 event. Aftershocks have the potential to cause further damage, especially to structures weakened by the initial large earthquake. The USGS noted that buildings in the Port-au-Prince area will continue to be at risk from strong earthquake shaking, and that the fault responsible for the January 12, 2010, earthquake still stores sufficient strain to be released as a large, damaging earthquake during the lifetime of structures built during the reconstruction effort.[26]

The USGS based its probability estimates on techniques developed to assess earthquake hazards in the United States. Using these techniques, the USGS estimated that the probability of a magnitude 7 or greater earthquake occurring within the next 50 years along the Enriquillo fault near Port-au-Prince is between 5% and 15%. The range of probabilities reflects the current understanding of the seismicity and tectonics of the Haiti region. By comparison, the USGS has estimated that that the probability of a magnitude 7 or greater earthquake occurring within the next 50 years along the Hayward-Rodgers Creek fault east of San Francisco is about 15%.[27]

February 27, 2010, Magnitude 8.8 Earthquake in Chile

A magnitude 8.8 earthquake struck Chile on February 27, 2010, along a subduction zone plate boundary fault 65 miles north-northeast of the city of Concepcion and offshore of the Chilean coast.[28] The earthquake occurred at a depth of approximately 22 miles below the seafloor, much deeper than the earthquake that struck Haiti on January 12, 2010. The city of Concepcion experienced intensity IX shaking on the Modified Mercalli Intensity Index, corresponding to considerable damage to specially designed structures, and corresponding to great damage to "substantial" buildings. The capital city of Santiago, located 200 miles northeast of the epicenter, experienced intensity VIII shaking corresponding to considerable damage in ordinary substantial buildings.[29] The earthquake caused an estimated $30 billion in total economic damage.[30] Over 500 deaths were reported, many from the tsunami generated by the subsea earthquake, and approximately 1.8 million people were affected.

Because the earthquake occurred offshore, it generated a tsunami, which struck parts of the Chilean coastline and offshore islands, causing damage and fatalities. Tsunami warnings were issued by the National Weather Service Pacific Tsunami Warning Center for Hawaii, Japan, and other regions bordering the Pacific Ocean that may have been vulnerable to a damaging tsunami wave, although most regions far from the epicenter did not experience any serious damage. A tsunami caused significant damage to the city of Hilo, Hawaii, following the May 1960 magnitude 9.5 earthquake that also occurred along the subduction zone fault about 143 miles south of the February 27, 2010, earthquake.[31] Why the 1960 earthquake generated a tsunami that caused damage and fatalities in Hawaii, Japan, and the Philippines, while the 2010 earthquake did not, is not yet well understood and is being actively studied.

The magnitude 8.8 earthquake occurred along the boundary between the Nazca tectonic plate and the South American tectonic plate, which converge at a rate of about 3 inches per year. The Nazca plate is subducting under the South American plate, which rides over the top of the Nazca plate. In geologic terms, this is known as a thrust fault or megathrust, in contrast to a strike-slip fault, where the rocks on either side of the fault slide past each other. The San Andreas fault and the Enriquillo fault that caused the January 2010 Haiti earthquake are strike-slip faults. The Sumatran-Andaman megathrust fault, which triggered the December 2004 Indonesian earthquake and tsunami, is a subduction zone fault or megathrust geologically similar to the Nazca-South American tectonic plate subduction zone.

March 11, 2011, Magnitude 9.0 Earthquake in Japan

A 9.0 magnitude massive earthquake struck off Japan's northeast coast near Honshu on March 11, 2011. The earthquake triggered a tsunami that caused widespread devastation to parts of the coastal regions in Japan closest to the earthquake epicenter. The epicenter was located about 80 miles east of Sendai, and about 230 miles northeast of Tokyo, and it occurred at a depth of approximately 20 miles beneath the seafloor.[32]

The earthquake resulted from thrust faulting along the subduction zone plate boundary between the Pacific and North America plates, and this is similar tectonically to the motion described for the 2010 Chile earthquake. Where the earthquake occurred, the Pacific plate is moving westward and sliding underneath the North America plate at just over 3 inches per year. (See *Figure 2.)*

This is similar to the convergence rate of the Nazca plate and the South American plate on the west side of Chile, where the February 27, 2010, earthquake occurred. The convergence zone between the Pacific plate and North America plate creates an undersea feature known as the Japan Trench. According to the USGS, tectonic plate motion in the Japan Trench subduction zone has triggered nine magnitude 7 or greater earthquakes since 1973.[33] Also, records indicate that large offshore earthquakes occurred in the same subduction zone in 1611, 1896, and 1933, each producing tsunamis that caused great destruction and fatalities.[34] According to records, the 1896 earthquake created tsunami waves of over 100 feet high and a reported death toll of 27,000.[35]

Is There a Similar Risk to the United States?

Subduction zone megathrust faults generate the largest earthquakes in the world. The Cascadia Subduction Zone megathrust that stretches from mid-Vancouver Island in southern British Columbia southward to Cape Mendocino in northern California has the potential to generate a very large earthquake, similar in magnitude to the February 2010 Chilean earthquake and the March 11, 2011 Japan earthquake.

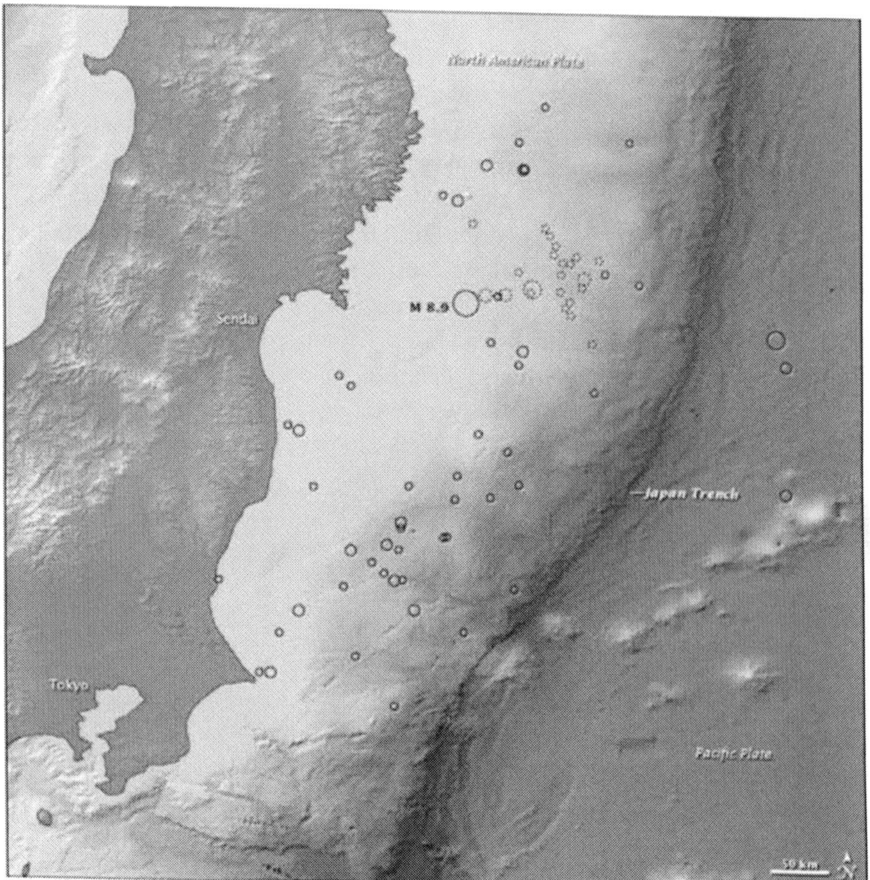

Source: NASA, Earth Observatory, March 11, 2011, http://earthobservatory.nasa.gov/NaturalHazards/view.php? id=49621. Notes: Large circle depicts epicenter of the earthquake (upgraded to magnitude 9.0); solid circles indicate aftershocks, dotted circles indicate foreshocks (smaller earthquakes that occurred prior to the major earthquake).

Figure 2. Image of the Japan Trench and Location of the March 11, 2011, Earthquake. (the Pacific plate is moving west and underneath the North America plate)

The fault's proximity to the northwestern U.S. coastline— approximately 50-100 miles offshore—also poses a significant tsunami hazard; destructive waves from a large earthquake along the fault could reach the coast of Oregon and Washington in less than an hour, possibly in tens of minutes. The Cascadia Subduction Zone fault forms the boundary between the subducting Juan de Fuca tectonic plate and the overriding North America plate, very similar to the relationship between the Nazca plate and the South American plate off the Chilean coast, and the Pacific plate and North American plate east of Japan. If the Cascadia Subduction Zone megathrust were to "unzip" or rupture along a large section of its entire length, models indicate that it would likely generate a megathrust earthquake near magnitude 9 or more, similar to the 1964 Alaskan earthquake, the 1960 and 2010 Chilean earthquakes, the 2004 Indonesian earthquake, and the 2011 Japan earthquake. Scientists have documented that the last time this occurred along the Cascadia Subduction Zone fault was in 1700. The 1700 earthquake spawned a tsunami that traveled across the Pacific Ocean and struck Japan. Because of the similarities in the subduction zone megathrust faults, scientists hope to learn a great deal about the seismic hazard in the Pacific Northwest by studying the unique strong ground motion recordings from the 2010 Chilean magnitude 8.8 earthquake and the 2011 Japan earthquake.

MONITORING

Congress authorized the USGS to monitor seismic activity in the United States in the 1990 law modifying NEHRP (P.L. 101-614).[36] The USGS operates a nationwide network of seismographic stations called the Advanced National Seismic System (ANSS), which includes the National Strong-Motion Project (NSMP). Globally, the USGS and the Incorporated Research Institutions for Seismology (IRIS) operate 140 seismic stations of the Global Seismic Network (GSN) in more than 80 countries.[37] The GSN provides worldwide coverage of earthquakes, including reporting and research.[38]

Advanced National Seismic System (ANSS)

According to the USGS, "the mission of ANSS is to provide accurate and timely data and information products for seismic events, including their effects on buildings and structures, employing modern monitoring methods and technologies."[39] If fully implemented, ANSS would encompass more than 7,000 earthquake sensor systems covering portions of the nation that are vulnerable to earthquake hazards.[40] As envisioned, the system would consist of dense urban networks, regional networks, and backbone stations.

Dense Urban Networks

In the original conception for ANSS, approximately 6,000 of the planned stations would have been installed in 26 high-risk urban areas to monitor strong ground shaking and how buildings and other structures respond. Currently, five high-risk urban areas have instruments deployed in sufficient density to generate the data to produce near real-time maps,[41] called ShakeMaps, which can be used in emergency response during and after an earthquake.[42] (See "ShakeMap," below.)

Backbone Stations

Approximately 100 instruments comprise the existing "backbone" of ANSS, with a roughly uniform distribution across the United States, including Alaska and Hawaii. These instruments provide a broad and uniform minimum threshold of coverage across the country. The backbone network consists of USGS-deployed instruments and other instruments that serve both ANSS and the EarthScope project (described below, under "National Science Foundation").

National Strong-Motion Project (NSMP)

Under ANSS, the USGS operates the NSMP to record seismic data from damaging earthquakes in the United States on the ground and in buildings and other structures in densely urbanized areas. The program currently has approximately 1,280 strong-motion[43] instruments across the United States and in the Caribbean. The NSMP has three components: data acquisition, data management, and research. The near real-time measurements collected by the NSMP are used by other government agencies for emergency response and real-time warnings. If fully implemented, the ANSS program would deploy about 3,000 strong-motion instruments. Many of the current NSMP instruments are older designs and are being upgraded with modern seismometers.

Regional Networks

If ANSS were fully implemented under its original conception, approximately 1,000 new instruments would replace aging and obsolete stations in the networks that now monitor the nation's most seismically active regions. The current regional networks contain a mix of modern, digital, broadband, and high-resolution instruments that can provide real-time data; they are supplemented by older instruments that may require manual downloading of data. Universities in the region typically operate the regional networks and will likely continue to do so as ANSS is implemented.

Global Seismic Network (GSN)

The GSN is a system of broadband digital seismographs around the globe, designed to collect high-quality data that are readily accessible to users worldwide, typically via computer. Currently, 140 stations have been installed in 80 countries and the system is nearly complete, although in some regions the spacing and location of stations has not fully met the original goal of uniform spacing of approximately 2,000 kilometers. The system is currently providing data to the United States and other countries and institutions for earthquake reporting and research, as well as for monitoring nuclear explosions to assess compliance with the Comprehensive Test Ban Treaty.

The Incorporated Research Institutions for Seismology (IRIS) coordinates the GSN and manages and makes available the large amounts of data that are generated from the network. The actual network of seismographs is organized into two main components, each managed separately. The USGS operates two-thirds of the stations from its Albuquerque Seismological Laboratory, and the University of California-San Diego manages the other third via its Project IDA (International Deployment of Accelerometers). Other universities and affiliated agencies and institutions operate a small number of additional stations. IRIS, with funding from the NSF, supports all of the stations not funded through the USGS appropriations. Funding for the GSN is provided via annual appropriations from the USGS and the National Science Foundation. In addition, the USGS committed $4.7 million from ARRA funding to the GSN, and NSF committed a similar portion of its ARRA funding to replace obsolete equipment on GSN stations worldwide.[44]

DETECTION, NOTIFICATION, AND WARNING

Unlike other natural hazards, such as hurricanes, where predicting the location and timing of landfall is becoming increasingly accurate, the scientific understanding of earthquakes does not yet allow for precise earthquake prediction. Instead, notification and warning typically involves communicating the location and magnitude of an earthquake as soon as possible after the event to emergency response providers and others who need the information.

Some probabilistic earthquake forecasts are now available that give, for example, a 24-hour probability of earthquake aftershocks for a particular region, such as California. These forecasts are not predictions, and are currently intended to increase public awareness of the seismic hazard, improve emergency response, and increase scientific understanding of the short-term hazard.[45] In the California example, a time-dependent map is created and updated every hour by a system that considers all earthquakes, large and small, detected by the California Integrated Seismic Network,[46] and calculates a probability that each earthquake will be followed by an aftershock[47] that can cause strong shaking. The probabilities are calculated from known behavior of aftershocks and the possible shaking pattern based on historical data.

When a destructive earthquake occurs in the United States or other countries, the first reports of its location, or epicenter,[48] and magnitude originate either from the NEIC or from the regional seismic networks that are part of ANSS. Other organizations, such as universities, consortia, and individual seismologists may also contribute information about the earthquake after the event. Products such as ShakeMap (described below) are assembled as rapidly as possible to assist in emergency response and damage estimation following a destructive earthquake.

National Earthquake Information Center (NEIC)

The NEIC, in Golden, CO, is part of the USGS. Originally established as part of the National Ocean Survey (U.S. Department of Commerce) in 1966, the NEIC was made part of the USGS in 1973. With data gathered from the networks described above and from other sources, the NEIC determines the location and size of all destructive earthquakes that occur worldwide and disseminates the information to the appropriate national or international agencies, government public information channels, news media, scientists and scientific groups, and the general public.

With the advent of the USGS Earthquake Notification Service (ENS), notifications of earthquakes detected by the ANSS/NEIC are provided free to interested parties. Users of the service can specify the regions of interest, establish notification thresholds of earthquake magnitude, designate whether they wish to receive notification of aftershocks, and even set different magnitude thresholds for daytime or nighttime to trigger a notification.

The NEIC has long-standing agreements with key emergency response groups, federal, state, and local authorities, and other key organizations in earthquake-prone regions who receive automated alerts—typically location and magnitude of an earthquake—within a few minutes of an event in the United States. The NEIC sends these preliminary alerts by email and pager immediately after an earthquake's magnitude and epicenter are automatically determined by computer. This initial determination is then checked by around-the-clock staff who confirm and update the magnitude and location data.[49] After the confirmation, a second set of notifications and confirmations are triggered to key recipients by email, pager, fax, and telephone.

For earthquakes outside the United States, the NEIC notifies the State Department Operations Center, and often sends alerts directly to staff at American embassies and consulates in the affected countries, to the International Red Cross, the U.N. Department of Humanitarian Affairs, and other recipients who have made arrangements to receive alerts.

ShakeMap

Traditionally, the information commonly available following a destructive earthquake has been epicenter and magnitude, as in the data provided by the NEIC described above. Those two parameters by themselves, however, do not always indicate the intensity of shaking and extent of damage following a major earthquake. Recently, the USGS developed a product called ShakeMap that provides a nearly real-time map of ground motion and shaking intensity following an earthquake in areas of the United States where the ShakeMap system is in place. *Figure 3* shows an example of a ShakeMap.

The maps produced portray the extent of damaging shaking and can be used by emergency response and for estimating loss following a major earthquake. Currently, ShakeMaps are available for northern California, southern California, the Pacific Northwest, Nevada, Utah, Hawaii, and Alaska.[50]

With improvements to the regional seismographic networks in the areas where ShakeMap is available, new real-time telemetry from the region, and advances in digital communication and computation, ShakeMaps are now triggered automatically and made available within minutes of the event via the web. In addition, better maps are now available because of recent improvements in understanding the relationship between the ground motions recorded during the earthquake and the intensity of resulting damage.

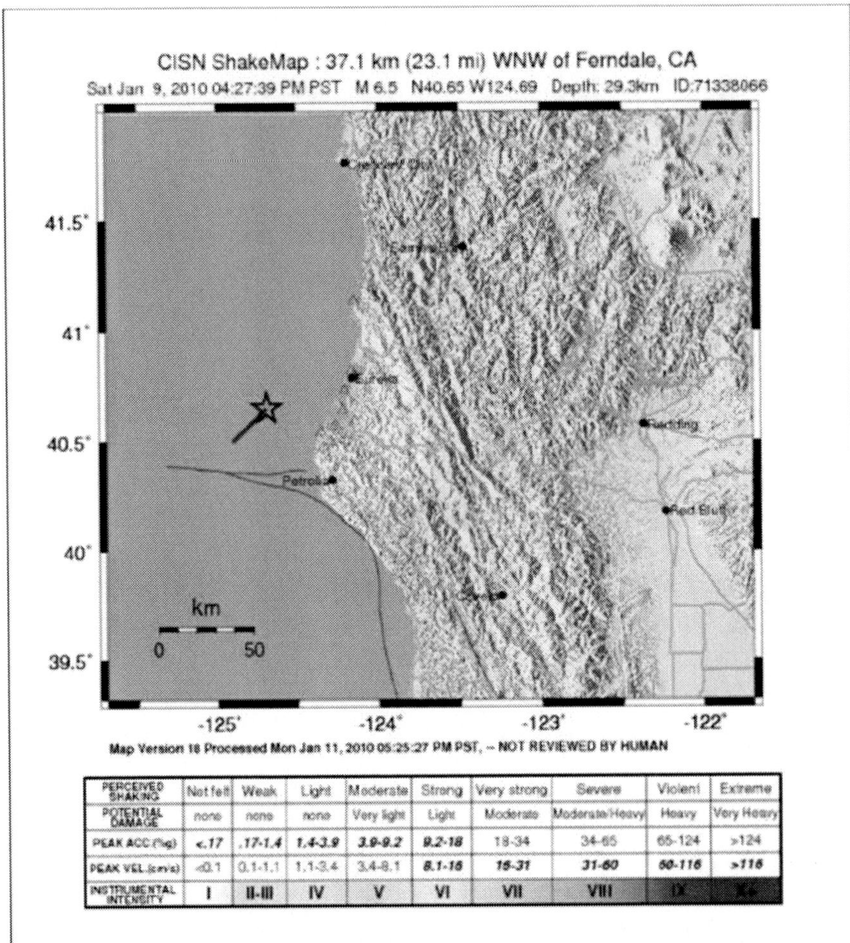

Source: USGS, http://earthquake.usgs.gov/eqcenter/shakemap/nc/shake/71338066/.

Note: Earthquake occurred 23.1 miles west-northwest of Ferndale, CA, at 4:27 p.m. on January 9, 2010, with a magnitude of 6.5. The star indicates the epicenter of the earthquake. Viewed on January 12, 2010.

Figure 3. Example of a ShakeMap.

If databases containing inventories of buildings and lifelines are available, they can be combined with shaking intensity data to produce maps of estimated damage. The ShakeMaps have limitations, especially during the first few minutes following an earthquake before additional data arrive from distributed sources. Because they are generated automatically, the initial maps are preliminary, and may not have been reviewed by experts when first made available. They are considered a work in progress, but are deemed to be very promising, especially as more modern seismic instruments are added to the regional networks under ANSS and computational and telecommunication abilities improve.

Prompt Assessment of Global Earthquakes for Response (PAGER)

Another USGS product that is designed to provide nearly real-time earthquake information to emergency responders, government agencies, and the media is the Prompt Assessment of Global Earthquakes for Response, or PAGER, system.[51] This automated system rapidly assesses the number of people, cities, and regions exposed to severe shaking by an earthquake, and generally makes results available within 30 minutes. Following the determination of earthquake location and magnitude, the PAGER system calculates the degree of ground shaking using the methodology developed for ShakeMap, estimates the number of people exposed to various levels of shaking, and produces a description of the vulnerability of the exposed population and infrastructure. The vulnerability includes potential for earthquake-triggered landslides, which could be devastating, as was the case for the huge May 12, 2008, earthquake in Sichuan, China. The automated and rapid reports produced by the PAGER system provide an advantage compared to the traditional accounts from eye-witnesses on the ground or media reports, because communications networks may have been disabled from the earthquake. Emergency responders, relief organizations, and government agencies could make plans based on PAGER system reports even before getting "ground-truth" information from eye-witnesses and the media.[52] *Figure 4* shows an example of PAGER output for the January 12, 2010, magnitude 7.0 earthquake in Haiti.

Pre-disaster Planning: HAZUS-MH

FEMA developed a methodology and software program called the Hazards U.S. Multi-Hazard (HAZUS-MH).[53] The program allows a user to estimate losses from damaging earthquakes, hurricane winds, and floods before a disaster occurs.

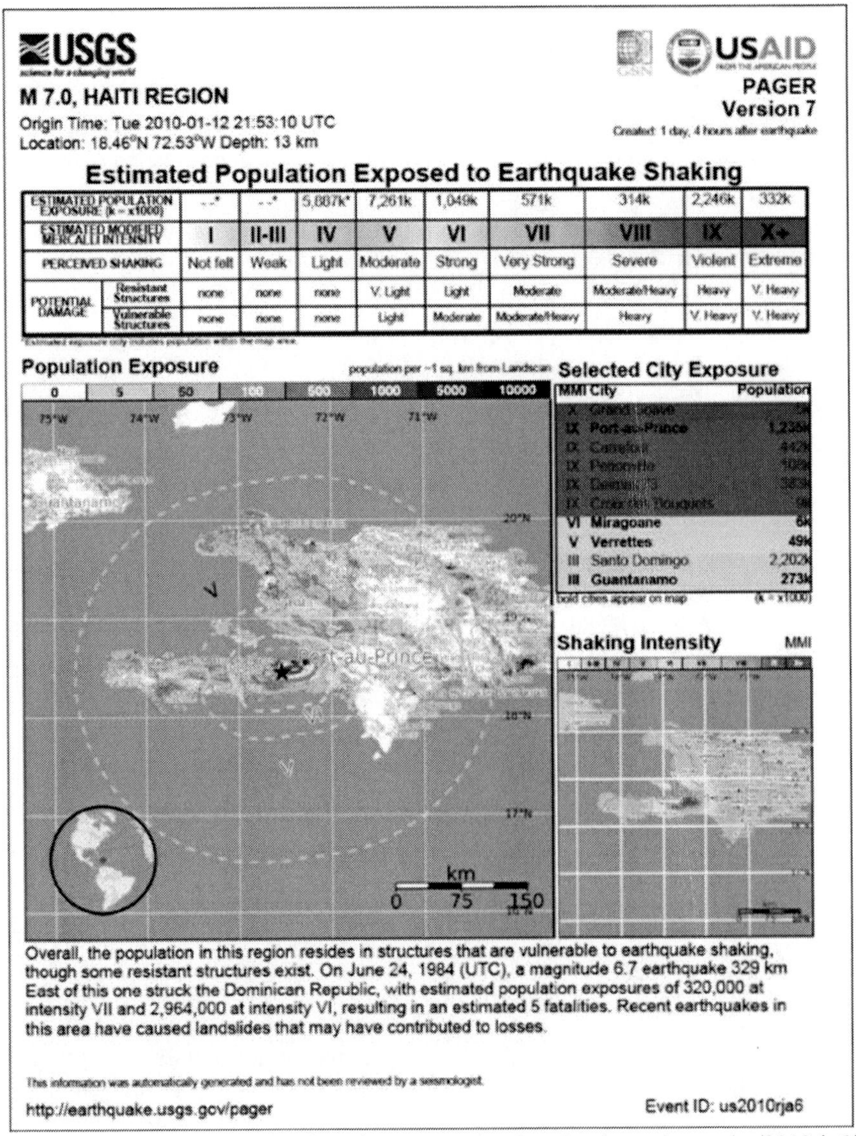

ESTIMATED POPULATION EXPOSURE (k = x1000)		--*	--*	5,887k*	7,261k	1,049k	571k	314k	2,246k	332k
ESTIMATED MODIFIED MERCALLI INTENSITY		I	II-III	IV	V	VI	VII	VIII	IX	X+
PERCEIVED SHAKING		Not felt	Weak	Light	Moderate	Strong	Very Strong	Severe	Violent	Extreme
POTENTIAL DAMAGE	Resistant Structures	none	none	none	V. Light	Light	Moderate	Moderate/Heavy	Heavy	V. Heavy
	Vulnerable Structures	none	none	none	Light	Moderate	Moderate/Heavy	Heavy	V. Heavy	V. Heavy

*Estimated exposure only includes population within the map area.

Source: USGS, http://earthquake.usgs.gov/earthquakes/pager/events/us/2010rja6/onepager.pdf.

Note: This is version 7 of the PAGER output, accessed on January 14, 2010.

Figure 4. Example of PAGER Output for the January 12, 2010, Magnitude 7.0 Haiti Earthquake.

The pre-disaster estimates could provide a basis for developing mitigation plans and policies, preparing for emergencies, and planning response and recovery. HAZUS-MH combines existing scientific knowledge about earthquakes (for example, ShakeMaps, described above), engineering information that includes data on how structures respond to shaking, and geographic information system (GIS) software to produce maps and display hazards data including economic loss estimates. The loss estimates produced by HAZUS-MH include:

- physical damage to residential and commercial buildings, schools, critical facilities, and infrastructure;
- economic loss, including lost jobs, business interruptions, repair and reconstruction costs; and
- social impacts, including estimates of shelter requirements, displaced households, and number of people exposed to the disaster.

In addition to furnishing information as part of earthquake mitigation efforts, HAZUS-MH can be used to support real-time emergency response activities by state and federal agencies after a disaster. Twenty-seven HAZUS-MH user groups—cooperative ventures among private, public, and academic organizations that use the HAZUS-MH software—have formed across the United States to help foster better-informed risk management for earthquakes and other natural hazards.[54]

RESEARCH—UNDERSTANDING EARTHQUAKES

U.S. Geological Survey

Under NEHRP, the USGS has responsibility for conducting targeted research into improving the basic scientific understanding of earthquake processes. The current earthquake research program at the USGS covers six broad categories:[55]

- *Borehole geophysics and rock mechanics*: studies to understand heat flow, stress, fluid pressure, and the mechanical behavior of fault-zone materials at seismogenic[56] depths to yield improved models of the earthquake cycle;

- *Crustal deformation*: studies of the distortion or deformation of the earth's surface near active faults as a result of the motion of tectonic plates;
- *Earthquake geology and paleoseismology*: studies of the history, effects, and mechanics of earthquakes;
- *Earthquake hazards*: studies of where, why, when, and how earthquakes occur;
- *Regional and whole-earth structure*: studies using seismic waves from earthquakes and man-made sources to determine the structure of the planet ranging from the local scale, to the whole crust, mantle, and even the earth's core; and
- *Strong-motion seismology, site response, and ground motion*: studies of large-amplitude ground motions and the response of engineered structures to those motions using accelerometers.

National Science Foundation

NSF supports fundamental research into understanding the earth's dynamic crust. Through its Earth Sciences Division (part of the Geosciences Directorate), NSF distributes research grants and coordinates programs investigating the crustal processes that lead to earthquakes around the globe.[57]

EarthScope

In 2003, NSF initiated a Major Research Equipment and Facilities Construction (MREFC) project called EarthScope that deploys instruments across the United States to study the structure and evolution of the North American Continent, and to investigate the physical processes that cause earthquakes and volcanic eruptions.[58] EarthScope is a multi-year project begun in 2003 that is funded by NSF and conducted in partnership with the USGS and NASA.

EarthScope instruments are intended to form a framework for broad, integrated studies of the four-dimensional (three spatial dimensions, plus time) structure of North America. The project is divided into three main programs:

- *The San Andreas Fault Observatory at Depth (SAFOD)*, a deep borehole observatory drilled through the San Andreas fault zone close to the hypocenter of the 1966 Parkfield, CA, magnitude 6 earthquake;

- *The Plate Boundary Observatory (PBO)*, a system of GPS arrays and strainmeters[59] that measure the active boundary zone between the Pacific and North American tectonic plates in the western United States; and
- *USArray*, 400 transportable seismometers that will be deployed systematically across the United States on a uniform grid to provide a complete image of North America from continuous seismic measurements.

SAFOD and PBO are in place and providing data to the seismological community. USArray is progressing across North America and is also furnishing real-time data to seismologists. The portable array has progressed across most of the conterminous United States. Remaining stations will be installed in New England, upstate New York, and southern Quebec. The installation plan calls for completing the portable array by the end of 2013.[60]

Network for Earthquake Engineering Simulation

Through its Engineering Directorate, NSF funds the George E. Brown Jr. Network for Earthquake Engineering Simulation (NEES), a project intended to operate until 2014, aimed at understanding the effects of earthquakes on structures and materials.[61] To achieve the program's goal, the NEES facilities conduct experiments and computer simulations of how buildings, bridges, utilities, coastal regions, and materials behave during an earthquake. In the first six years of operations since 2004, 160 multiyear projects have been completed or are in progress under NEES.[62]

OUTLOOK

A precise relationship between earthquake mitigation measures, NEHRP and other federal earthquake-related activities, such as earthquake research, and reduced losses from an actual earthquake may never be possible. However, as more accurate seismic hazard maps evolve, as understanding of the relationship between ground motion and building safety improves, and as new tools for issuing warnings and alerts such as ShakeMap and PAGER are devised, trends denoting the effectiveness of mitigation strategies and earthquake research and other activities may emerge more clearly. Without an ability to precisely predict earthquakes, Congress is likely to face an ongoing

challenge in determining the most effective federal approach to increasing the nation's resilience to low-probability but high-impact major earthquakes.

End Notes

[1] The NEHRP program is discussed in CRS Report R43141, The National Earthquake Hazards Reduction Program (NEHRP): Issues in Brief, by Peter Folger.

[2] U.S. Geological Survey, Department of the Interior, Earthquake Hazards—A National Threat, Fact Sheet 2006-3016, March 2006, http://pubs.usgs.gov/fs/2006/3016/2006-3016.pdf. During the period 1975-1995, only four states did not experience detectable earthquakes: Florida, Iowa, North Dakota, and Wisconsin. See USGS Earthquake Hazards Program, Earthquake Facts, at http://earthquake.usgs.gov/learn/facts.php.

[3] USGS Fact Sheet 2008-3018, "2008 United States National Seismic Hazard Maps" (April 2008), at http://pubs.usgs.gov/fs/2008/3018/pdf/FS08-3018_508.pdf;

[4] USGS Fact Sheet 2008-3027, "Forecasting California's Earthquakes—What Can We Expect in the Next 30 Years?" (2008), at http://pubs.usgs.gov/fs/2008/3027/fs2008-3027.pdf.

[5] U.S. Geological Survey, Earthquakes with 1,000 or More Deaths Since 1900, at http://earthquake.usgs.gov/ earthquakes/world/world_deaths.php. This estimate does not include fatalities from the February 27, 2010, magnitude 8.8 Chilean earthquake, which has resulted in widespread destruction but few fatalities compared to the Indonesian, Pakistan, and Haiti earthquakes.

[6] Estimates of earthquake-related fatalities vary, and an exact tally of deaths and injuries is rare. For more information on the difficulties of counting earthquake-related deaths and injuries, see http://earthquake.usgs.gov/regional/world/ casualty_totals.php.

[7] FEMA Publication 366, HAZUS MH Estimated Annualized Earthquake Losses for the United States (April 2008), at http://www.fema.gov/library/viewRecord.do?id=3265. Hereafter referred to as FEMA 366.

[8] A. M. Best Company Inc., 2006 Annual Earthquake Study: $100 Billion of Insured Loss in 40 Seconds (Oldwick, NJ: A.M. Best Company, 2006), p. 12. The A. M. Best report includes estimates from catastrophe-modeling companies of predicted damage from hypothetical earthquakes in Los Angeles, the Midwest, the Northeast, and Japan. The report cites an estimate by one such company, Risk Management Solutions (RMS), that a hypothetical 7.4 magnitude event along the Newport-Inglewood Fault near Los Angeles would cause $549 billion in total property damage. A hypothetical 6.5 magnitude earthquake along a fault between Philadelphia and New York City would produce $901 billion in total loss, according to an RMS estimate.

[9] Kevin Vranes and Roger Pielke, Jr., "Normalized Earthquake Damage and Fatalities in the United States: 1900- 2005," Natural Hazards Review, vol. 10, no. 3 (August 2009), pp. 84-101.

[10] Building inventory refers to four main inventory groups: (1) general building stock, (2) essential and high potential loss facilities, (3) transportation systems, and (4) utility systems (FEMA 366).

[11] FEMA 366, p. 37.

[12] FEMA 366, p. 10.

[13] The largest earthquakes in New York, New Jersey, and Massachusetts were, respectively: 1944, Massena, NY, magnitude 5.8, felt from Canada south to Maryland; 1783, New Jersey,

magnitude 5.3, felt from New Hampshire to Pennsylvania; and 1755, Cape Ann and Boston, MA, intensity of VIII on the Modified Mercalli Scale, felt from Nova Scotia to Chesapeake Bay (USGS Earthquake Hazards Program).

[14] USGS Circular 1188, Table 3.

[15] FEMA 366, p. 32.

[16] Ibid., p. 32 and p. 36.

[17] Ibid., p. 36.

[18] In contrast to California, where earthquakes occur on the active margin of the North American tectonic plate, the New Madrid seismic zone is not on a plate boundary but may be related to old faults in the interior of the plate, marking a zone of tectonic weakness.

[19] Andrew Newman et al., "Slow Deformation and Lower Seismic Hazard in the New Madrid Seismic Zone," Science, v. 284 (April 23, 1999), pp. 619-621.

[20] Seth Stein, Joseph Tomasello, and Andrew Newman, "Should Memphis Build for California's Earthquakes?" Eos, v. 84, no. 19, (May 13, 2003), pp. 177, 184-185.

[21] Seth Stein, "Code Red: Earthquake Imminent?" Earth, vol. 54, no. 1 (January 2009), pp. 52-59.

[22] Some researchers measure, for example, less than 2 millimeters of ground motion per year in the New Madrid Seismic Zone using modern GPS technology. In contrast, motion across the San Andreas Fault in California is about 36 millimeters per year. See Seth Stein, Disaster Deferred: How New Science is Changing Our View of Earthquake Hazards in the Midwest (New York: Columbia University Press, 2010), pp. 4-5.

[23] According to the USGS, the May 22, 1960, magnitude 9.5 earthquake was the largest earthquake in the world. See http://earthquake.usgs.gov/earthquakes/world/ events/ 1960_05_22.php.

[24] USGS Earthquake Hazards Program, at http://earthquake.usgs.gov/earthquakes/eqinthenews/ 2010/us2010rja6/.

[25] See http://earthquake.usgs.gov/earthquakes/eqinthenews/2010/us2010rja6/#summary.

[26] USGS statement, "USGS Updates Assessment of Earthquake Hazard and Safety in Haiti and the Caribbean," February 23, 2010, at http://www.usgs.gov/newsroom/article.asp? ID=2413&from=rss_home.

[27] Ibid. However, the USGS also notes that the probability of a magnitude 6.7 or greater earthquake occurring on the Hayward-Rodgers fault over the next 30 years is 31%.

[28] See http://earthquake.usgs.gov/earthquakes/eqinthenews/2010/us2010tfan/#details.

[29] See http://earthquake.usgs.gov/earthquakes/eqinthenews/2010/us2010tfan/#summary.

[30] Ibid.

[31] The Orphan Tsunami of 1700—Japanese Clues to a Parent Earthquake in North America, USGS, Professional Paper 1707, 2005, http://pubs.usgs.gov/pp/pp1707/.

[32] USGS, Earthquake Hazards Program, http://earthquake.usgs.gov/earthquakes/eqinthenews/ 2011/usc0001xgp/.

[33] USGS Earthquake Hazards Program, http://earthquake.usgs.gov/earthquakes/eqinthenews/ 2011/usc0001xgp/#summary.

[34] Ibid.

[35] For more information on the March 11, 2011, Japan tsunami, and the U.S. tsunami monitoring network, see CRS Report R41686, U.S. Tsunami Programs: A Brief Overview, by Peter Folger.

[36] For more information about NEHRP, see CRS Report R43141, The National Earthquake Hazards Reduction Program (NEHRP): Issues in Brief, by Peter Folger.

37 IRIS is a university research consortium, primarily funded by NSF, that collects and distributes seismographic data.

[38] The GSN also monitors nuclear explosions.

[39] USGS Earthquake Hazards Program, http://earthquake.usgs.gov/research/monitoring/anss/.

[40] According to the USGS, the ANSS system is about 30% complete, with 2,564 ANSS stations installed at the end of 2012. USGS FY2014 budget justification, p. H-9.

[41] The five urban areas are Los Angeles, San Francisco, Seattle, Salt Lake City, and Anchorage. E-mail from William Leith, USGS, February 7, 2011.

[42] The number of stations necessary to generate a data-based ShakeMap depends on the urban area and geology, but roughly correspond to about half the number of planned stations per urban area, at a spacing of about 20 kilometers between stations. Personal communication, William Leith, USGS, January 11, 2010.

[43] Strong motion seismometers, or accelerometers, are special sensors that measure the acceleration of the ground during large (>6.0 magnitude) earthquakes.

[44] USGS FY2011 Budget Justification, p. J-32. The USGS portion of annual appropriations in FY2013 was $5.2 million.

[45] USGS Open-File Report 2004-1390, and California 24-hour Aftershock Forecast Map, at http://pasadena.wr.usgs.gov/step/.

[46] The California Integrated Seismic Network is the California region of ANSS; see http://www.cisn.org/.

[47] Earthquakes typically occur in clusters, in which the earthquake with the largest magnitude is called the main shock, events before the main shock are called foreshocks, and those after are called aftershocks. See also http://pasadena.wr.usgs.gov/step/aftershocks.html.

[48] The epicenter of an earthquake is the point on the earth's surface directly above the hypocenter. The hypocenter is the location beneath the earth's surface where the fault rupture begins.

[49] In early 2006, the NEIC implemented an around-the-clock operation center and seismic event processing center in response to the Indonesian earthquake and resulting tsunami of December 2004. Funding to implement 24/7 operations was provided by P.L. 109-13.

[50] ShakeMaps for some areas outside the United States are also available. See http://earthquake.usgs.gov/eqcenter/ shakemap/.

[51] See the USGS Earthquakes Hazards Program for more information, at http://earthquake. usgs.gov/earthquakes/pager/.

[52] See also USGS Fact Sheet 2007-3101 at http://pubs.usgs.gov/fs/2007/3101/.

[53] See http://www.fema.gov/plan/prevent/hazus/hz_overview.shtm.

[54] See http://www.hazus.org/.

[55] See http://earthquake.usgs.gov/research/.

[56] Seismogenic means capable of generating earthquakes.

[57] See http://www.nsf.gov/div/index.jsp?div=EAR.

[58] See http://www.earthscope.org/.

[59] A strainmeter is a tool used by seismologists to measure the motion of one point relative to another.

[60] See http://www.usarray.org/maps.

[61] Management for NEES has been headquartered at Purdue University's Discovery Park since October 1, 2009. Institutions participating in NEES include Cornell University; Lehigh University; Oregon State University; Rensselaer Polytechnical Institute; University of Buffalo-State University of New York; University of California-Berkeley; University of California-Davis; University of California-Los Angeles; University of California-San Diego; University of California-Santa Barbara; University of Colorado-Boulder; University

of Illinois at Urbana-Champaign; University of Minnesota; University of Nevada-Reno; and University of Texas at Austin. See http://www.nees.org/.

[62] See http://nees.org/about.

INDEX

A

B

C

S

T

U